HETEROTYPICAL BEHAVIOUR IN MAN AND ANIMALS

HETEROTYPICAL BEHAVIOUR IN MAN AND ANIMALS

Edited by

M. HAUG

Psychology Laboratory,
Louis Pasteur University,
Strasbourg

P.F. BRAIN,

Biological Sciences,
University College of
Swansea

and

C. ARON

Histology Institute,
Faculty of Medicine,
Strasbourg

CHAPMAN AND HALL

London ● New York ● Tokyo ● Melbourne ● Madras

UK   Chapman and Hall, 2-6 Boundary Row, London SE1 8HN

USA   Chapman and Hall, 29 West 35th Street, New York NY10001

JAPAN   Chapman and Hall Japan, Thomson Publishing Japan,
            Hirakawacho Nemoto Building, 7F, 1-7-11 Hirakawa-cho,
            Chiyoda-ku,
            Tokyo 102

AUSTRALIA   Chapman and Hall Australia, Thomas Nelson Australia,
            480 La Trobe Street, PO Box 4725, Melbourne 3000

INDIA   Chapman and Hall India, R. Seshadri, 32 Second Main
            Road, CIT East, Madras 600 035

First edition 1991

© 1991 M. Haug, P. Brain and C. Aron

Printed in Great Britain by
T.J. Press (Padstow) Ltd, Padstow, Cornwall

ISBN 0 412 33260 4

British Library Cataloguing in Publication Data
Heterotypical behaviour in man and animals.
1. Sexual behaviour. Animals 2. Man. Sexual behaviour
I. Haug, Marc II. Brian, Paul F. III. Aron, Claude
155.8.
ISBN 0-412-33260-4

Library of Congress Cataloguing-in-Publication Data
Heterotypical behaviour in man and animals/edited by M. Haug, P.F. Brain. and C.
Aron – 1st ed.
p. cm.
Includes bibliographical references and index.
ISBN 0-412-33260-4
1. Sex (Biology) 2. Sexual behaviour in animals. 3. Aggressive behaviour in animals.
4. Physiology, Comparative. 5. Sex (Psychology)—Endocrine aspects. I. Haug, M.
(Marc) II. Brain, Paul F. III. Aron. Claude.
[DNLM: 1. Aggressin. 2. Sex behaviour. 3. Sex behaviour, Animal. 4. Sex
Hormones—physiology. HQ 23 H589]
QP251.H46 1981
599.056—dc20
DNLM/DLC
for Library of Congress

Contents

Acknowledgements

We thank gratefully Marie Claire Hantsch and Jean Marrel for their invaluable help with respect to the production of this book.

Preface

Etienne E. Baulieu*

The theme of this book, Heterotypical Behaviour in Man and Animals, should be of great interest to physiologists, endocrinologists, physicians, and workers in social sciences. Although Heterotypical <u>Sexual</u> Behaviour is a major theme, this volume attempts to display wide interest in reproductive medicine, general physiology, and behaviour in the two sexes. The editors explore the psycho-social dimension, not only of sexuality, but of eroticism which, as recalled by John Money, has its etymological root in the Greek word for love. Being an endocrinologist, who has studied hormone function in terms of synthesis, metabolism, distribution and receptors of these messenger molecules, I would like to recall some data which are basic when considering the overall human machine.

It is common knowledge that androgens and oestrogens are formed in both sexes, differences being observed only in concentrations and rhythms of secretion. In the brain of the two sexes, there appear to be the same enzymes which may transform androgens to oestrogens, a process which could explain some aspects of CNS differentiation and activity. Both males and females have androgen and oestrogen receptors, and neurally these receptors appear to be present at the same order of magnitude and distributed according to the same pattern. There is even a similar distribution of receptors for progesterone, the hormone of pregnancy, in the brains of males and females. Therefore, several important pieces of the machinery transmitting sexual information

---

* Laureat of the 1989 Albert Lasker Clinical Medical Research Award.
INSERM U33, Département de Chimie Biologique, UER Kremlin-Bicêtre, 78 Avenue du Général Leclerc, 94270 Bicêtre, France

are similar or identical, and probably do not explain "homotypical" or "heterotypical" behaviours. Some surprises may, however, be evident in the next few years if the techniques of molecular genetics are applied to these problems. Much will be learned concerning the characteristics of the target cells of hormones and how these characteristics are acquired during differentiation.

There is still a great deal to learn about hormones per se. The recent finding of de novo steroid formation from cholesterol by glial cells (oligodendrocytes) - apparently similar in male and female (rats), and probably also in human beings - indicates that it is not sufficient to consider blood concentration of hormones synthesized in the peripheral organs if one wants to understand cerebral functions. It also seems to me very important to distinguish between activity induced by an excess, a defect, or a given pulse of a hormone, expressed by the classic dose-to-effect relationship, and the permissive role of certain hormones, particularly glucocorticosteroids. For permissive activities one needs the presence of a hormone at a certain level, but more of it does not modify the cellular response in all cells. The concepts of classical and permissive hormones may correspond to a sort of double information system in all cells : at the nuclear genetic level (e.g. by steroid hormones) and at the surface (membrane) level (e.g. by peptidic hormones). A third general concept that must also to be considered is the multipotentiality of hormones which can be engaged in differentiation processes as well as in the control of cellular activities. A hormone's effect may, for example, depend on the age of the subject, other hormones, or on the presence of oncogenes. One should not forget that the state of neural activity is also related to action, emotion and thinking.

It is therefore very difficult to discern in a given individual what features are influenced by hormonal or nonhormonal processes and which are consequences of recent or past events. Even limited indices of deviance may be of great interest, for instance when considering the formation of homosexual and transexual patterns of behaviour. It is also important to work out a definition of the matter under study, considering the obvious intervention of social components. I favour the enlarging of the expression Sex Typical Behaviour to Sex Conforming Behaviour, and even more appropriately to Gender Typical Behaviour. I would, however, like to see an endocrine approach not only of sex determined characteristics, but of gender associated criteria.

Much interesting work is still to come. This well edited book will contribute to attempts to define priorities.

1

Are Behaviours Specific to Animals of Particular Sex ?

*Paul F. Brain * and Marc Haug ***

The processes accounting for sex differences in form and behaviour in mammals are said to be due to subtle interactions between genes, hormones and neural circuits acting at different times of the organism's life. As, however, heterotypical behaviour is relatively common (especially in female vertebrates) one has to ask the questions (a) is such behaviour simply erroneous ; (b) is the idea that behaviours can be sexually dimorphic fundamentally flawed, or (c) are there circumstances when behaviour typical of the "other" sex can be advantageous to particular animals ? This initial account will briefly review the data on the presumed biological origins of sexual dimorphic behaviours and then will examine the topics treated in the rest of this diverse volume.

Effects of Genes

Byskov (1986) suggested that morphological sex differentiation and the inflexibility of phenotypic sex are basic mammalian characteristics. Indeed, gonadal sex in such species is largely a consequence of genetic sex which is permanently fixed at the moment of gestation. Morphological sex of an embryo is generally determined by its gonadal sex. If a testis differentiates, male development occurs, and if an ovary differentiates, female development ensues. Subsequently, the time-scale of production and number of gonocytes differ in male and female mammals (for reviews see Dillon, 1973 ; Short, 1979 ; Steinberger, 1971).

---

* Biological Sciences, University College of Swansea, Swansea, SA2 8PP, U.K.

** Université Louis Pasteur, Laboratoire de Psychophysiologie, (URA 1295) 7 rue de l'Université, 67000 Strasbourg, France

Sexual differentiation commences in mammals during the early foetal period (Jost, 1972). In such vertebrates, males are the heterogametic (or XY) and females the homogametic (or XX) sex. With the exception of minor histocompatibility genes (e.g. the H-Y antigen), the Y chromosome does not seem to possess any structural genes except those for initiating (directly or indirectly) the differentiation of the male gonads. Half of the Y chromosome is made up of repeated DNA sequences, lacking obvious genetic functions but which have no counterpart in female cells. Actually, their function may be to increase the size of the Y chromosome and prevent its pairing (during meiosis) with the unmatched X chromosome.

The discovery of the sex chromatin "Barr body" in the nuclei of female cells (Moore & Barr, 1953) suggested that one X chromosome in that sex is normally inactivated. This process may be clinically extended, to a degree, to karyotypes with supernumerary X chromosomes. XXX females have two Barr bodies whereas XXY and XXXY males have one and two such entities, respectively. XO females, like normal XY males, have no Barr body. In spite of the apparently superfluous nature of one X chromosome in normal females, the gonadal and somatic characteristics of XO humans (Homo sapiens) includes amenorrhoea (failure to menstruate), short stature and a webbed neck. This suggests that the second X chromosome in humans serves some, as yet unidentified, function. In contrast, XO mice (Mus musculus) are not necessarily sterile. The effect of the XXY karyotype in humans (Homo sapiens) is variable : some individuals are normal and fertile whereas others are underdeveloped sexually, aggressive and/or mentally subnormal (Mittwoch, 1967 ; Simpson, 1976).

The faster growth rate of the testis may reflect the shorter period of DNA synthesis in the male, due to differences in initial DNA content of XY and XX cells (Mittwoch, 1970, 1973) or to the self-stimulatory effect on growth of the developing organ's early testosterone (T) secretion.

Effects of Early Hormone Exposure

Many investigators have observed that the variations in aggressiveness between different strains of rodents may reflect hormonally-mediated effects during foetal life rather than differences in the frequencies of genes (Svare & al., 1984). Steroid hormones require, however, the presence of hormone receptors in target tissues to express their actions. Variations in phenotype may thus be due to differences in steroid receptor concentrations rather than to hormonal titre per se. For example, the testicular feminization syndrome seems due to an X-linked mutation which results in a decreased production of androgen receptors in genetic males. Such males secrete T but are unable to respond to it normally, resulting in a failure of masculinization during foetal life. This phenomenon has been described by Fox & al. (1982) in many species including humans, mice and rats (Rattus norvegicus).

Male humans have significantly higher blood titres of T during the foetal period of sexual differentiation than their female counterparts. In contrast, female foetuses have higher blood titres of oestradiol (E2) than do males (Belisle & Tulchinsky, 1980). An essentially similar pattern of sexually dimorphic foetal steroid titres occurs in mice (Vom Saal & Bronson, 1980).

"Masculinization" is the process of acquiring male-typical characteristics and "defeminization" is a loss of female-typical features. Payne and Swanson (1972) have pointed out that in golden hamsters (Mesocricetus auratus) these processes are, to some extent, independent. Defeminization is not simply the converse of masculinization. In rodents with short gestation periods (e.g. mice, rats and hamsters) exposure of female neonates to androgen both masculinizes (Christensen & Gorski, 1978) and defeminizes (Manning & McGill, 1974) their subsequent behaviour. Conversely, castration of male neonates demasculinizes (Beach & al., 1969) and feminizes (Corbier & al., 1983) adult behaviour. The female offspring of pregnant female rats injected on post-conception day 19 with the pharmacological dose of 1mg testosterone propionate (TP) are defeminized (Huffman & Hendricks, 1981).

Effects of Dimorphic Brain Nuclei

In rats, the perinatal sex steroid pattern produces morphological differences in various parts of the hypothalamus which have been described by numerous authors (Arai & Kusama, 1968 ; Balazs, 1974 ; Dörner & Staudt, 1968 ; Harris, 1970 ; Pfaff, 1966 ; Raisman & Field, 1973, etc.)

In squirrel monkeys (Saimiri sciureus), the cell nuclei of the male are significantly larger than those of females in the nucleus medialis amygdalae (Bubenik & Brown, 1973). Cells in the suprachiasmatic nucleus (SCN), medial POA, the arcuate nucleus and the cortex are, however, the same size in the sexes. Differences also appear, however, in the distribution of terminals of non-amygdaloid fibres in the POA and in the lateral nucleus of the amygdala in rats (Balazs, 1974 ; Raisman & Field, 1973). These morphological differences reflect functional changes, for example, in the sensitivity of these neural loci to sex steroids. In later life, there is a correlation between the size of the above sexually-dimorphic brain nuclei and the pattern of sexual behaviour (Hutt, 1972).

Breedlove & Arnold (1981) have provided evidence that neonatal steroids can influence neuronal size, and Toran-Allerand (1984) demonstrated that steroids can influence neurite growth in neonatal tissue in rats. The size of such nuclei in males, is also increased after neonatal castration.

Prenatal exposure of pregnant rats to antiandrogens (e.g. 5mg/day of flutamide from day 10 after conception to parturition) demasculinizes both male and female offspring (Clemens & al., 1978). In contrast, implantation of T or E2 pellets into the POA-VMH area of the brains of female rats, produces masculinization of subsequent behaviour (Christensen & Gorski, 1978).

Neonatal androgenization of female rats reduces E2 binding in oestrogen-sensitive tissue (Flerko & Mess, 1968 ; Maurer & Woolley, 1974 ; Vertes & King, 1971). Moreover, the deleterious effects of oestrogen (e.g. producing abnormal sexual differentiation) are not confined to the brain. For example, the synthetic oestrogen RU2858 induces reproductive tract abnormalities and increases the incidence of primary vaginal carcinoma in female human children exposed to this drug in foetal life (Bongiovanni & al., 1959 ; Harbest & al., 1972). Although understanding of the morphological sex differences in the CNS seems currently rather poor in the mammals (Moore, 1985), it does seem to be a potentially important area of investigation.

Effects of Hormone Titre in Adulthood

Sexually dimorphic behaviours also depend on the hormones that an animal secretes in post-pubertal life. For example, untreated female mice will attack females in a fashion similar to that seen in inter-male aggression. Exposure to T, however, leads such animals to also attack male intruders (Brain & al., 1983 ; DeBold & Miczek, 1981). A number of recent studies have shown that adult female rats, when treated with TP, show levels of aggression which are comparable to those of males (Van de Poll & al., 1981a ; 1981b ; 1985). In contrast to animal studies, clinical investigations have used oestrogen to control "aggressive" behaviour in man (reviewed in Brain, 1984). In such studies, oestrogen is thought to block the action of the individual's endogenous T. This suggests that the failure of adult female rodents to secrete T, accounts for some of their differences from males in terms of agonistic behaviour.

Many studies in rodents also show that adult titres of gonadal hormones play roles in the development and maintenance of sexual behaviour. Female rats display feminine sexual behaviour (e.g. lordosis) which depends on the presence of gonadal steroids (Feder, 1978). Clemens & al. (1978) and Hasan (1987) found that ovariectomized female rats will show masculine sexual behaviour (e.g. mounting) after T treatment. Some studies suggest, in contrast, that T treatment of female rats produces receptive or proceptive behaviour (Aron & Asch, 1963 ; McDonald & Meyerson, 1973 ; Meyerson & al., 1979). Female rats prefer the company of males when they are in the proestrus or oestrus phase of their cycles (Eliasson & Meyerson, 1975) or when they are ovariectomized and exogenously-treated with either oestradiol benzoate

(EB) or TP (De Jonge & al., 1986a ; McDonald & Meyerson, 1973 ; Meyerson & Lindstrom, 1973 ; Meyerson & al., 1979). Additional treatment with progesterone did not, however, increase the female's orientation preference towards sexually active males (De Jonge & al., 1986b). It has even been suggested that androgens are normally involved in the sexual motivation or "libido" of female rats (De Jonge & Van de Poll, 1982), monkeys (Everitt & Herbert, 1972 ; Nadler & al., 1983) and humans (Gorzynski & Katz, 1977 ; Bancroft, 1980).

TOPICS IN THE PRESENT VOLUME

Hurtig and Pichevin (this volume) essentially open our volume by discussing sex typicality and sex conformity, pointing out that, in order to investigate heterotypical behaviour, one first has to define sex-typical behaviour. They illustrate many of the problems in this field asking, for example, whether sex differences are always arranged in the same manner. In particular, they discuss the postulate of psychological dimorphism which maintains that there are attitudes and behaviours associated with the sex categories. Hurtig and Pichevin discuss the ambiguity of the expression "sex differences" and debate the varied meanings of the term "sex typical" to be found in the literature. Essentially, they extend the review of research on sex differences by Maccoby and Jacklin (1974) by taking issue with the claim of these earlier reviewers that there are some "well-established" sex differences in the literature. Hurtig and Pichevin claim that research on sex differences shows a wide range of applications and internal fragility (i.e. the findings are generally unreliable and unstable). Indeed, there seem to be problems with the behavioural indices used and definitions of what to study. They conclude that, if the legitimacy of the constructs has not been established, it is "premature and misleading" to talk of group differences. Hurtig and Pichevin deal with the "well-established" sex differences in verbal, visual-spatial and mathematical abilities together with aggression in our species, examining the appropriateness of the meta-analyses. They conclude that even in the strongest of these "well-established" differences (visual-spatial ability and aggressive behaviour) that sex accounts for no more than 5% of total variation. An important point is that within sex variation is greater than between sex variation. Hurtig and Pichevin advocate the current approach of regarding masculinity and femininity as independent dimensions and introduce the notion of psychological androgyny. They suggest that using the term sex conformity has advantages and that describing behaviours as being gender-typical is better than talking of sex differences.

Vom Saal (this volume) then essentially looks at some of the recent complexities revealed by studies on mice and rats. He notes that the basic assumption that masculinization and defeminization of males is an active process mediated by a variety of hormones (including androgens and their metabolites) whereas the female phenotype is generated by a passive process (see introduction) has been contradicted by recent data. Some "masculine" and "feminine" behaviours may be changed in opposite directions by the same manipulations. Vom Saal notes that the sources of testosterone in rodent foetuses include the placenta as well as the testis. He notes, indeed, that the placenta of the female may secrete more androgen than that of the male. He naturally concentrates on the intrauterine location phenomenon where subjects are exposed to different amounts of testosterone by virtue of the sex of their neighbours within the uterus. Vom Saal records that females do not respond to the same degree as males to the same titre of testosterone. He suggests that there are essentially two approaches to studying the impact of exposure of female foetuses to testosterone and their subsequent "masculinization". One approach uses females from different intrauterine locations treated with testosterone in adulthood to induce heterotypical behaviour. He notes that neonatal anti-androgens reduce the effects of adult testosterone on such animals. There seems, however, to be a dearth of studies on female typical behaviour in such animals. Vom Saal also reviews the second approach which considers the potential effects of different intrauterine positions on reproductive success in natural habitats.

Vom Saal notes that studies that have and have not employed perinatal pharmacological manipulations have led to rather different conclusions concerning the roles of sex steroids in development of "normal" male and female phenotypes in rodents. OM females (developing adjacent to no male siblings) show a reproductive advantage only when the population density is low. 2M females (developing between two males) are more aggressive than OM counterparts and may be at an advantage under other circumstances. Vom Saal records that OM females are exposed to more oestradiol and less testosterone than their 2M counterparts. Males are also influenced behaviourally by their intrauterine locations and 2M males are more responsive to testosterone than their OM counterparts. Vom Saal emphasises that the hormone dynamics are complex involving actions at a variety of levels, metabolic conversions and the impact of binding globulins. The differentiation of the "masculine" or the "feminine" phenotype even in the "simple" rodent is consequently much more complex than was proposed in the initial model.

Goy and Roy (this volume) look at mounting as the most common heterotypical behaviour in _female_ mammals (rodents, carnivores and primates). They note that many females show more mounting in oestrus and ask whether ovarian hormones play a role in this behaviour. They note that rodents show considerable variation but record that hormonal factors in early development change the potential for showing mounting. For example, prenatal oestrogens and androgens have an impact on such behaviour in guinea pigs. Goy and Roy note that in some species (strains ?) biological conversion of testosterone to oestradiol is important but that this is not the case for primates (oestrogens administered before birth do not cause psychosexual virilisation in primates). Goy and Roy also review the evidence for there being sexually dimorphic nuclei in the brains of mammals and the effects of lesioning such areas.

These authors identify three broad classes of mounting by normal females, namely that which is (a) free of ovarian influences and persists after ovariectomy ; (b) an integral part of oestrus behaviour and is related to hormonal changes, and (c) induced by treatments with large doses of sex steroids and is normally only seen at a low incidence in females. Goy and Roy note that not all species display all these classes and suggest that the broad types may be regulated by different neuronal populations. They further note that categories of female mounting can be divided into "true heterotypical" (associated with masculinization) and "quasi heterotypical" (independent of masculinization) varieties. In terms of the functions of heterotypical behaviour in females, they record that roles of such activities have been claimed in play, aggression and sexual behaviour. One intriguing suggestion is that such behaviour may be a form of reciprocal altruism serving to attract a male to an individual related to the mounter. It has also been suggested that female mounting can be employed to arouse a sluggish mate.

Aron & al. (this volume) state that, despite the organizational effects of androgens in the perinatal period, males of some mammalian species have the capacity to display feminine behaviour in adulthood. In their extensive studies on rats, Aron & al. noted that progesterone was unable to increase the lordosis response in oestrogen-primed males, but that, by contrast, orchidectomized animals were not only capable of displaying lordosis following oestrogen treatment but also proved to be sensitive to the facilitatory effects of exogenous or endogenous progesterone. They also report that, in the absence of any hormonal or surgical manipulation, male rats may display strain-dependent heterotypic sexual responses to mounts by other males. Consequently, the organizational effects of androgens on male sexual behaviour must be considered as being impermanent. Aron & al. also demonstrate that a common behavioural system sensitive to the activational effects of ovarian hormones and which involves both the hypothalamic ventromedial nucleus and the different amygdaloid nuclei subserve the display of lordosis in male as well as in female rats. They found that, in contrast to unexposed animals, castrated rats primed with ovarian hormones (oestradiol and progesterone) show lordosis more frequently in response to appropriate male odours.

However, the main objective of Aron & al. was to evaluate, in the light of experimental and physiopathological data, the respective roles of hormonal, environmental and organizational factors in the determinism of sexual orientation in humans. Human data are reviewed on 5α-reductase deficiency in pseudohermaphrodites where individuals are said to change their gender identity and role at puberty. One should note that Money (this volume) and Götz & al. (this volume) interpret these data differently. Aron & al. suggest that, under favourable cultural conditions, the perinatal organizational effects of androgen may overcome the influences of environmental factors. They caution however that one should not underestimate the potency of environment influences, making reference to the number of females masculinized at birth by the congenital adrenal hyperplasia condition, who go on to be normal females after corrective surgery.

The main question raised by Aron & al. is to what extent the use of an animal model may lead to a better understanding of human sexuality. They lay emphasis on the fact that sexual orientation in rodents has been conclusively shown to depend upon a balance between organizational and environmental factors. In the absence of any other evidence, why then disregard the biological background of behavioural bisexuality in humans ?

Money (this volume) suggests that sex has a broad reputation, probably as a result of the conflicts between reproductive biologists and sexologists. He suggests that there is primarily a conflict between reductionist and multivariate complexity types of explanation and reiterates that the distinction between biological and sociocultural explanations is inherently sterile. Whilst recognizing their usefulness for certain purposes, Money feels that animal models for human erotosexualism are inadequate.

As noted earlier, Money has considerable reservations about the 5α-reductase deficiency/pseudohermaphrodism story which essentially suggests that this enzyme system determines "male sex drive" and "male gender identity". He suggests that the culture in which such individuals are found makes it easier for them to claim to be males at puberty.

In considering gender identity and gender role, Money discusses the roles of prenatal hormones, especially as revealed in genetic abnormalities. He suggests that, in humans, progesterone may act as an antivirilizing agent in the female foetus in contrast to the impact of foetoprotein and sex steroids in rodents. In terms of masculinization and feminization, Money favours the ambitypic model which suggests that the activation of one (e.g. masculinization) does not reciprocally deactivate the other (e.g. defeminization). Money suggests that one can identify sex-shared/threshold-dimorphic behaviours in animal and clinical models. These include kinetic energy expenditure, roaming, competitive rivalry/assertiveness in childhood, fighting predators, fighting in defence of rehearsal and responses to visual or tactile erotic stimuli. Money feels that neonatal bonding has an important impact on erotosexual development, a feature perhaps over-emphasized by Freud. Processes like identification and reciprocation are involved. The author notes that flirtaceous and erotosexual rehearsal play are likely to occur and is concerned about the impact of preventing this on adult sexual behaviour. Money clearly believes that the idea that sexual behaviour occurs in our species only at puberty (the sexual latency phenomenon) is a myth.

In considerations of homosexual, bisexual and heterosexual predispositions, Money does not feel it inappropriate to use the term limerence (the state of being a person who falls in love) for all of them. He suggests that the lovemap (a template in the brain of males and females of the "perfect" lover/love affair) differs. Money also discusses how there may be disorders of sub-processes in sexual behaviour e.g. in proception, acception and conception in females. He also points out that subjects such as erotosexual cyclicity and erotosexual gerontology have received remarkably little attention.

Gotz & al. (this volume) describe the differentiation of the gonads and the secondary sexual organs in mammals. They also relate the steps of sexual differentiation in our own species, claiming that the ratio of androgen to oestrogen is responsible for sexual differentiation of the brain. These authors discuss homosexuality and trans-sexuality and their potential hormonal bases. Götz & al. describe the sex specific patterns of pituitary gonadotrophin secretion in a range of mammals and maintain that oestrogen effects (particularly in early life) are important. They especially studied the modified oestrogen feedback effects on gonadotrophins in populations of homo-, hetero- and trans-sexual men and women. Men with the highest luteinizing hormone levels four days after oestradiol treatment show the lowest testosterone titres four days after treatment with human chorionic gonadotro-phin. Homosexuals seem to evidence changed gonadotrophin responsiveness.

Götz & al. also consider the claim (also mentioned by Aron & al.) of a positive association between stress and homosexuality. They review the data on individuals apparently exposed to increased stress in World War II, in individuals with varied exposure to stress in early development (largely due to problems with the maternal-infant bond) and the impact of stress on animal studies. Their basic conclusion is that stress increases the inci-dence of homosexuality. There are, however, interpretational prob-lems in such studies (see Whalen, this volume). Another intriguing claim is that another enzyme deficiency (in 21-hydroxylase) may be implicated in female-to-male trans-sexuals.

Langevin (this volume) concentrates his attention on the varied topic of aggression where it has been repeatedly claimed that males are more aggressive than females. He reviews background information on gun ownership and homicide, prior aggressive experiences and homicide (narcissism is often evident in sex killers), alcohol and drug abuse and situational strains (e.g. "love triangles"). Langevin also examines the potential roles of biological factors in such phenomena including brain damage, sex hormones (especially testosterone), physical illness (e.g. diabetes) and genetic factors. It can be concluded that this "sexually dimorphic" trait is really rather difficult to deal with, being influenced by a wide range of factors and being essentially a rather heterogeneous phenomenon.

Whalen (this volume) finally provides an overview (which draws on his vast accumulated experience with hormones and behaviour) for the volume. He reiterates that it has become increasingly apparent that gonadal hormones often exert their effects after metabolic conversion. Whalen notes that (in his experience) particular species and strains (of mice) may respond differently to metabolites of testosterone. Naturally, this genetic variance makes the extraction of basic principles difficult if not impossible. He suggests that analogical rather than homological thinking is more likely to advance our understanding of behaviour in mice and humans.

Whalen notes that, in a volume in which sex differences are a focus, it is important to record that intrasex variability is at least as large as intersex variability. In spite of attempts to relate "sexually dimorphic" behaviours to biological factors, it is essential to record that behaviour is much influenced by the social setting in which the organism finds itself. Whalen interestingly opines that traditional statistical treatments have tended to mask or preclude considerations of the sources of variance. This author clearly recognizes that factors (such as hormones) can influence behaviour via a variety of means including the production of social signals and the perception of those factors but suggests that the brain is the primary "organ of behaviour". He provides a useful account of the history of the discovery of sexual dimorphism in the brain but suggests that we still find it difficult to know which sexually dimorphic structures should be related to which kinds of behaviour. Whalen airs a number of models and strategies which may prove useful in this respect.

Clearly, Heterotypical Behaviour in Man and Animals can only, at this stage, provide a flavour of the nature of the debates developing within an area characterised by considerable problems and ethical difficulties. In spite of the differences between several of the accounts, it is obvious to everyone that the earlier ideas on what constitutes "sexually dimorphic behaviour" and how such behaviour is generated were gross oversimplifications. To answer the question posed in the title of this chapter, it seems highly improbable that any behaviour is uniquely specific to any particular sex. One can talk in terms of "sex-typical" or "gender-typical" behaviour for a particular strain of a particular species. One has to recognize, however, that other organisms under other environments may show very different patterns. This should not be too surprising given the adaptiveness of behaviour and the increased appreciation of the flexibility of a wide range of species. It is certainly inappropriate to regard an activity that appears to be heterotypical as being necessarily aberrant or abnormal. The complexities revealed in the varied accounts should also make us very wary of any attempt to extrapolate from a particular animal to the human condition. Having said this, a surprisingly large number of the potential complicating factors involved in studies on heterotypical behaviours have been characterised in rats, mice or infra-human primates.

REFERENCES

Arai, Y. and Kusama, T. (1968). Effect of neonatal treatment with estrone on hypothalamic neurons and regulation of gonadotrophin secretion. Neuroendocrinology, 3, 107-117.

Aron, C. and Asch, G. (1963). Action exercée par la testosterone au cours même du cycle oestral, sur le comportement sexuel et sur l'activité ovarienne chez la ratte. Comptes Rendus de la Société de Biologie, 157, 645-648.

Aron, C., Chateau, D. and Schaeffer, C. (1990). Heterotypic sexual behaviour in male mammals : the rat as an experimental model. This volume.

Balazs, R. (1974). Influence of metabolic factors on brain development. British Medical Bulletin, 30, 126-134.

Bancroft, J. (1980). Endocrinology of sexual function. Clinical Obstetrics and Gynaecology, 7, 253-280.

Beach, F.A., Noble, R.G. and Orndoff, R.K. (1969). Effect of perinatal androgen treatment on responses of male rats to gonadal hormones in adulthood. Journal of Comparative and Physiological Psychology, 68, 490-497.

Belisle, S. and Tulchinsky, D. (1980). Amniotic fluid hormones, In : Tulchinsky, D. & Ryan, K. (Eds) : Maternal-fetal Endocrinology (p. 169). W.B. Saunders & Co. Philadelphia.

Bongiovanni, A.M., DiGeorge, A.M. and Grumbach, M.M. (1959). Masculinization of female infant associated with estrogen therapy alone during gestation : Four cases. Journal of Clinical Endocrinology and Metabolism, 19, 1004-1011.

Brain, P.F. (1984). Biological explanations of human aggression and the resulting therapies offered by such approaches : a critical evaluation. In : Blanchard, R.J. & Blanchard, D.C. (Eds). Advances in the Study of Aggression Volume 1 (pp. 63-102). New York : Academic Press.

Brain, P.F., Haug, M. and Alias bin Kamis (1983). Hormones and different tests for "aggression" with particular reference to the effects of testosterone metabolites. In : Balthazart, J., Prove, E. & Gilles, R. (Eds), Hormones and Behaviour in Higher Vertebrates. (pp. 290-304). Berlin : Springer-Verlag.

Breedlove, N. and Arnold, A. (1981). Sexually dimorphic motor nucleus in rat spinal cord : response to adult hormone manipulation, absence in androgen insensitive rats. Brain Research, 225, 297-307.

Bubenik, G.A. and Brown, G.M. (1973). Morphologic sex differences in the primate brain areas involved in regulation of reproductive activity. Experientia, 15, 619-621.

Byskov, A.G. (1986). Differentiation of the mammalian embryonic gonad. Physiological Reviews, 66, 71-117.

Christensen, L.C. and Gorski, R. (1978). Independent masculinization of neuroendocrine system by intracerebral implants of testosterone or estradiol in the neonatal female rat. Brain Research, 146, 325-340.

Clemens, L.G., Gladue, B.A. and Coniglio, L.P. (1978). Prenatal endogenous androgenic influences on masculine sexual behavior and genital morphology in male and female rats. Hormones and Behavior, 10, 40-53.

Corbier, P., Roffi, J. and Rhoda, J. (1983). Female sexual behavior in male rats : effect of hour of castration at birth. Physiology and Behavior, 30, 613-616.

DeBold, J.F. and Miczek, K.A. (1981). Sexual dimorphisms in the hormonal control of aggressive behavior of rats. Pharmacology, Biochemistry and Behavior, 14, 89-93.

DeJonge, F.H. and Van de Poll, N.E. (1982). Sexual motivation and proceptive behavior in the rat. 4th Meeting of the European Society for Comparative Physiology and Biochemistry, Abstr. pp. 254-255.

DeJonge, F.H., Burger, J. and Van de Poll, N.E. (1986a). Acute effects of gonadal hormones on lordosis behavior of female rat. Brain Research, 20, 57-62.

DeJonge, F.H., Eerland, E.M.T. and Van de Poll, N.E. (1986b). The influence of oestrogen, testosterone and progesterone on partner preference, receptivity and proceptivity. Physiology and Behavior, 37, 885-891.

Dillon, R.S. (1973). Handbook of Endocrinology. Philadelphia : Lea and Febiger Publishers.

Dörner, G. and Staudt, J. (1968). Structural changes in the preoptic anterior hypothalamic area of the male rat, following neonatal castration and androgen substitution. Neuroendocrinology, 3, 136-140.

Eliasson, M. and Meyerson, B.J. (1975). Sexual preference in female rats during oestrous cycle, pregnancy and lactation. Physiology and Behavior, 14, 705-710.

Everitt, B.J. and Herbert, J. (1972). Hormonal correlates of sexual behaviour in subhuman primates. Danish Medical Bulletin, 19, 246-258.

Feder, H.H. (1978). Specificity of steroid hormone activation of sexual behavior in rodents. In : Hutchison, J.B. (Ed.), Biological Determinants of Sexual Behavior, (pp. 395-424). New-York : John Wiley and Sons Ltd.

Flerko, B. and Mess, B. (1968). Reduced oestradiol-binding capacity of androgen sterilized rats. Acta Physiologica Academia Scientiarum Hungaricae, 33, 111-113.

Fox, T., Olsen, K., Vito, C. and Wieland, S. (1982). Putative steroid receptors : genetics and development. In : Schmitt J. and Bloom, F. (Eds). Molecular Genetics and Neuroscience A New Hybrid (p. 289). Raven Press, New York.

Gorzynski, G. and Katz, J.L. (1977). The poly-cyclic ovary syndrome : psychosexual correlates. Archives of Sexual Behavior, 6, 215-218.

Götz, F., Rohde, W. and Dorner, G. (1990). Neuroendocrine diffe-rentiation of sex-specific gonadotrophin secretion, sexual orientation and gender role behaviour. This volume.

Goy, R.W. and Roy, M. (1990). Heterotypic sexual behaviour in female mammals. This volume.

Harbest, A.L., Kurman, R.J., Scully, R.E. and Poskanzen, D.C. (1972). Clear-cell adrenocarcinoma of the genital tract in young females. New England Journal of Medicine, 287, 1259-1264.

Harris, G.W. (1970). Hormonal differentiation of the developing central nervous system with respect to patterns of endocrine function. Philosophical Transactions of the Royal Society of London, 259, 165-177.

Hasan, S.A. (1987). Videotape Analysis of the Involvement of Sex Steroids in the Social interactions in Laboratory Mice and Rats. Ph.D. Dissertation University of Wales.

Huffman, L. and Hendricks, S.E. (1981). Prenatal injected testos-terone propionate and sexual behaviour of female rats. Physiology and Behavior, 28, 773-778.

Hurtig, M.C. and Pichevin, M.E. (1990). Sex typicality and sex conformity. This volume.

Hutt, C. (1972). Neuroendocrinological, behavioural and intellec-tual aspects of sexual differentiation in human development. In : Ounsted, C. and Taylor, D.C. (Eds). Gender Differences, (pp. 73-121). Edinburgh, Churchill-Livingstone.

Jost, A. (1972). A new look at the mechanisms controlling sex differentiation in mammals. Johns Hopkins Medical Journal, 130, 38-53.

Langevin, R. (1990). Biological and psychological factors in human aggression. This volume.

Maccoby, E.E. and Jacklin, C.N. (1974). The Psychology of Sex Differences, Stanford, CA : Stanford University Press.

Manning, A. and McGill, T.E. (1974). Neonatal androgen and sexual behaviour in female house mice. Hormones and Behavior, 5, 19-31.

Maurer, R.A. and Woolley, D.E. (1974). Demonstration of nuclear 3H-estradiol binding in hypothalamus and amygdala of female androgenized female and male rats. Neuroendocrinology, 16, 137-147.

McDonald, P.G. and Meyerson, B.J. (1973). The effect of oestradiol, testosterone and dihydrotestosterone on sexual motivation in the ovariectomized female rat. Physiology and Behavior, 11, 515-520.

Meyerson, B.J. and Lindstrom, L.H. (1973). Sexual motivation in the female rat. A methodological study applied to the investigation of the effects of oestradiol benzoate. Acta Physiologica Scandinavica Supplement, 389, 1-80.

Meyerson, B.J., Eliasson, M. and Hetta, J. (1979). Sex specific orientation manipulation. In : Kaye, A.M. (Ed). Advances in Bioscience, Volume 25. (pp. 451-460). Oxford : Pergamon Press.

Mittwoch, U. (1967). Sex chromosomes. New York : Academic Press.

Mittwoch, U. (1970). How does the Y chromosome affect gonadal differentiation ? Philosophical Transactions of the Royal Society of London, B 259, 113-117.

Mittwoch, U. (1973). Genetics of Sex Differentiation : New York, Academic Press.

Money, J. (1990). The development of sexuality and eroticism in human kind. This volume.

Moore, C.L. (1985). Another psychobiological view of sexual diffe-rentiation. Developmental Reviews, 5, 18-55.

Moore, K.L. and Barr, M.L. (1953). Morphology of the nerve cell nuclei with special reference to the sex chromatin. Journal of Comparative Neurology, 98, 213-231.

Nadler, R.D., Collines, D.C., Cherly, M.L. and Graham, Ch.E. (1983). Menstrual cycle patterns of hormones and sexual behavior of gorillas. Hormones and Behavior, 17, 1-18.

Payne, A.P. and Swanson, H.H. (1972). Neonatal androgenization of aggression in male golden hamster. Nature (London)., 239, 282-283.

Pfaff, D.W. (1966). Morphological changes in the brains of adult male rats after neonatal castration. Journal of Endocrinology, 36, 415-416.

Raisman, G. and Field, P.M. (1973). Sexual dimorphism in the neuropil of the preoptic area of the rat and its dependence on neonatal androgen. Brain Research, 54, 1-29.

Short, R.V. (1979). Sex determination and differentiation. British Medical Bulletin, 35, 121-127.

Simpson, J.L. (1976). Disorders of Sexual Differentiation. New York : Academic Press.

Steinberger, E. (1971). Hormonal control of mammalian spermatogenesis. Physiological Reviews, 51, 1-22.

Svare, B., Kinsley, C., Mann, M. and Broida, J. (1984). Infanticide accounting for genetic variation in mice. Physiology and Behavior, 33, 137-152.

Toran-Allerand, C.D. (1984). On the genesis of sexual differentiation of the central nervous system : morphogenetic consequences of steroidal exposure and possible role of   - foetoprotein. In : G.J. De Vries & al. (Eds). Progress in Brain Research. Volume 61 (pp. 63-98). Amsterdam : Elsevier Scientific Publishers B.V.

Van de Poll, N.E., De Jonge, F., Van Oyen, H.G., Van Pelt, J. and DeBruin, J.P.C. (1981a). Failure to find sex differences in testosterone-activated aggression in two strains of rats. Hormones and Behavior, 15, 94-105.

Van de Poll, N.E., Swanson, H.H. and Van Oyen, H.G. (1981b). Gonadal hormones and sex differences in aggression in rats. In : Brain P.F. and Benton, D. (Eds). The Biology of Aggression (pp. 243-252). Alphen Sijthoff and Noordhoff.

Van de Poll, N.E., Bowden, N.J., Van Oyen, H.G., De Jonge, F.H. and Swanson, H.H. (1985). Gonadal hormonal influences upon aggressive behaviour in male and female rats. In : Segal, M. (Ed). Psychopharmacology of Sexual Disorders. (pp. 63-77). London. John Libbey & Co.

Vertes, M. and King, R.J.B. (1971). The mechanism of oestradiol binding in rat hypothalamus : effect of androgenization. Journal of Endocrinology, 51, 271-282.

Vom Saal, F.S. (1990). Prenatal gonadal influences on mouse sociosexual behaviours. This volume.

Vom Saal, F. and Bronson, F. (1980). Sexual characteristics of adult female mice are correlated with their blood testosterone levels during prenatal development. Science (New York), 208, 597-599.

Whalen, R.E. (1990). Heterotypical behaviour in animals and man : concepts and strategies. This volume.

2
Sex Typicality and Sex Conformity

Marie-Claude Hurtig and Marie-France Pichevin *

It may seem paradoxical in a book on heterotypical behaviour to include a chapter on the complementary question "What is sex-typical behaviour ?". Yet it follows that in order for a given type of behaviour to be categorized as heterotypical, it must necessarily have been observed and identified as being characteristic of one - and not the other - of the sexes just as behaviour qualified as sexual is characteristic of one and only one of the sexes. However, some sexual behaviours of both sexes relate to the biological reproduction system. In other words, their emergence, form, and purpose depend on the requirements and constraints of that system, which in the human species happens to be biparental. Unlike the uniparental reproduction mode of other species, biparental reproduction requires specific behavioural patterns, such as those leading to the necessary meeting between two partners of the opposite sex. Although sexual dimorphism appears absolutely essential for the term "heterotypical" to make sense, and sexual dimorphism is found in conjunction with sexual behaviours that can be qualified as male or female, this does not mean that the wide variety of behaviour so-called sex-typical behaviour, dealt with in studies of human societies, has this basis.

Indeed, anthropologists, ethnologists, and sociologists have attempted to describe and analyze the different behaviours specific to each sex category, behaviours which make up the so-called sex roles. Research has shown that sex role behaviour is highly variable and flexible, both in ·space and across time. Indeed the modes of differentiation that affect the status of each of the sexes, the relationships between them, the distribution of power,

* Centre de Recherche en Psychologie Cognitive (CREPCO), Unité Associée au CNRS (UA 182), Université de Provence, 29 avenue Robert Schuman, 13621 Aix-en-Provence Cedex 1, France.
Paper translated from French by Vivian Lamongie.

the sexual division of work, etc... are also highly diverse. Although the above studies have been based on sexual dimorphism, they have nonetheless revealed just how culturally polymorphic this sexual dimorphism is.

Another apparent paradox here is that we should attempt to "academically" define a notion which, in the end, is so common to everyday psychology. "Another woman driver !", or "That's a man's job", or "Don't cry now - you're not a little girl after all" ; such statements have a familiar ring, and are all part of our social cognition of both the sex-typical and the heterotypical, being accessible, shared, and unquestioned. Let us take a closer look at this "common knowledge". It involves the capability to perceive, identify, and attribute characteristics that are not biological as sex-related. Thus, behaviour, professions, emotions, interpersonal relationships, body language, gestures, know-how, etc..., may all "belong" either to the realm of masculinity or femininity. Such knowledge even allows us to determine whether or not what an actor thinks, says, or does is appropriate to his/her sex. It expresses the atypical, prohibited, out of place, as well as the consistent, prescribed, expected, as if generating the "sex-appropriate" behaviour or the "right" response requires us to have to assure and reassure ourselves that our behaviour fits into the "system". This system establishes and is established by the relationship between our own or other's sex membership and the characteristics of the behaviour, a relationship which appears to guarantee sex identity, be a judgement grid, and a guide for action.

TYPICALITY AND DIMORPHISM

Sexual behaviour, sex role behaviour and sexed behaviour are respectively privileged links to the biological, the social and cultural and the cognitive and psychological. This delimits several disciplinary areas, where a number of research objects are built, and specific reference systems are called upon. Is there only one reality behind the single word "sex" ? Is the typical always defined in the same way when referring to sex ? Is dimorphism its best descriptor, and sex the unique architect ?

The Contribution of the Psychology of Sex Differences

It is the psychology of sex differences that will be investigated here in order to approach the above questions. This is for several reasons, namely : -
1) Because by outlining that which is sex-typical, we are not only describing the particular characteristics of each of the sexes, but determining what differentiates the two. The typical accentuates the different. Yet scientific psychology speaks mostly about differences between the sexes. There is a substantial amount of data in this field of research, making a

"finer" analysis of the dimensions of sex typicality possible.
If, indeed, this notion is to be considered as scientifically
relevant.

2) Because the study of how the social roles typical of each sex
are learned (i.e. of sex typing) focuses on the psychological
and social sex conformity processes, considered to be the
crucible of sex identity. Indeed, in the course of socialization,
each individual learns to differentiate the two sexes and,
through this difference, builds his/her individual and social
maleness or femaleness "according to whether he or she
shares the characteristics usually displayed by males or
females. This meaning of the term sex typing renders it
entirely redundant with the topic sex differences" (Maccoby &
Jacklin, 1974).

3) Because the notion of psychological dimorphism (so long the
central postulate of the psychology of sex differences) states
that "characteristics usually displayed by males or females"
do indeed exist, which in the study of personality have been
described as femininity and masculinity by going from an
analysis of behaviours to a psychology of traits. Sex-specific
psychological configurations then define the sexual types that
both make up identity styles, and are a reflection of the
standard, either natural or cultural (depending on the
theoretical standpoint).

4) Finally, because by the distinguishing between sex and gender,
naive psychology has gradually been taken into account by
considering the cognitive activity carried out when processing
information about sex (Bem, 1981). In this perspective, the
psychology of sex differences can provide us with some of the
elements of a theory of sex conformity which do not oppose
sex stereotypes - with their two distinct (male and female)
spheres of behaviour or traits. Real behaviour, when analyzed
scientifically, appears void of differentiation by sex.

Psychology and Dimorphism

Let us consider the postulate of psychological dimorphism
which states that there are behaviours, responses, attitudes,
personality traits, etc... whose emergence, form, and variations
are associated with the sex categories. It is obvious that the
nature and validity of this association are the determining factors
of the acceptance of this postulate. Evaluation of this relationship
ought to precede analysis of the characteristics of such
behaviours, responses, attitudes..., which in turn should contribute
to defining sex typicality. This postulate states, then, that the
difference between the two sex categories is basic. But it does
not say anything about intra-category (intragroup) differences or
inter-category (intergroup) similarities, nor does it deal with the
origins of such differences. Now, in both empirical research and

literature on the subject, (1) only the differences between the two sex categories have been taken into account, even though variations within a given category, as well as inter-sex similarities, are just as essential to defining the typical, and (2) heuristic value has been attributed to biological dimorphism, whether that attributing was supported by a biological theory or not. Biological dimorphism can be considered as a model or generator of the sexual dimorphism that can be extended to cover all human activities, some of whose forms then take on the role of secondary sexual characteristics, regardless of the levels of integration of the behaviour being studied, be it response patterns tightly linked to physiological specificities, such as breast feeding, or complex behaviour linked to sex role or social place, such as child care and child rearing by women.

The very ambiguity of the expression "sex differences" is a reflection of these notional slips. Morgan (1980) suggested that we only talk about "sex differences" when referring to secondary sexual characteristics in the strictest sense. His analysis is particularly relevant in our opinion to understanding the expression "sex-typical". This can be viewed as a corollary to what are usually called "sex differences" in the confusion between sex differences and sex-related differences.

Did you say "Typical" ?

Let us stop here for a moment and discuss the meaning of the term "typical", if we applied Morgan's distinction. Does saying that a trait, a behaviour, a response... is sex-typical mean that it is sexually dimorphic, or only that sex is a factor which has some effect on the variations observed in it ? According to the first acceptation of the word typical, which we will call the "hard" meaning :

1) The members of one of the two sex categories can be distinguished from members of the other by one or more traits, cognitive skills, social behaviours, etc..., which are thus characteristic of that sex and only that sex. Here is a typical descriptor of the sex-specific. It is definitional.

2) The relationship between an individual of a given sex and the presence or absence of one of these characteristic features is so close that it grants sex the predictive function. In other words, if the sex of a person is known, the presence or absence of a particular characteristic can be inferred with a high degree of accuracy, and inversely, the presence or absence of that characteristic in a person enables one to predict his/her sex. The typical, in this sense, operates as a means of inferring and identifying.

3) On the basis of this association (whether made by observation or prejudice) between the characteristic features and the sex of the actors, the features themselves can, by extension, be described as sexed, regardless of whether an actor is physi-

cally present or not. They can then be defined and classified
per se as male or female. The typical here operates as a
classifier, thus also defining the heterotypical.

Now let us take a look at the second acceptation of the term
"typical", which we will call the "soft" meaning :

1) Human traits, behaviours, attitudes, and responses are all
highly variable. Their variability, ascribable to social and
cultural factors, has above all been attributed to individual
factors. Using sex as a "natural" factor for grouping
individuals has (like using age) two advantages : (a) It
channels the variability by concentrating the data on the sex
categories and organizies it around the mean values for each
of the groups thus defined, and (b) when a statistically significant difference between the means is observed, it provides
an explanation of the variability by emphasising one of the
variation factors. The mean values thus represent two
constructed, standard forms around which individual variations
fluctuate within a given sex category. The differences - in
content, extent, and direction - between these means give
meaning to the difference between the two sex categories.
We might note in passing that the variable nature of human
behaviour and traits, theorized by Darwin, was itself
considered by psychologists and psychometricians in the early
20th-century as a sex characteristic, the greater variability of
males distinguishing them from females. "This one
fundamental difference in variability is more important than
all the differences between the average male and female
capacities" (Thorndike, 1906, quoted by Shields, 1975). This
quote may be compared to what Stanley Hall wrote in 1904,
"An ideal or typical male is hard to define but there is a
standard ideal woman". Today, Hall's assertion could be
interpreted in terms of prototypicality by granting standard
forms the status of prototypes. They serve as reference points
for measuring deviation from the norm. The typical here is
defined as that which complies with the standard, in a space
of behaviours or traits whose structure is not trivial. Is this
space structured as a continuum organized along two
differentiating standard forms, or is it made up of two
disjoint classes ? If we simply consider the difference
between the means, this question is unanswered.

2) Sex does not a priori have a predictive function. Maccoby and
Jacklin (1974) went so far as to question the validity of
grouping by sex by comparing this method to randomly
dividing the subjects into two groups. "The conclusion of their
evaluation of the research literature on sex differences seems
to suggest that assigning variables to groups by sex is
somewhat more meaningful than assigning them at random"
(Unger, 1979). What is the basis of this "meaningfulness" ?

The psychology of sex differences has often gone all too
quickly from sex-related differences to sex differences, from

the "soft" version to the "hard" version i.e. from the proto-typical to the typical, from a grouping variable to a causal factor.

WHAT DIFFERENCES ?

An important synthesis and review of previous research on sex differences was published in 1974 by Maccoby and Jacklin : The Psychology of Sex Differences. Their objective was to analyze the findings on sex differences and their variations with age, and to do so before attempting "to understand the 'why' and the 'how' of psychological sex differentiation. Since then, many other reviews have been published, most of which focus on a particular area of psychology, and either reanalyze the studies cited by Maccoby and Jacklin, or examine new data. We can even say that there has been a boom in work on this subject, partly due to the development of meta-analysis procedures (we will discuss this point in greater detail further on). We might mention reviews on cognitive skills (Fairweather, 1976 ; Hyde, 1981), visual-spatial ability (Caplan & al., 1985), conformity (Cooper, 1979) or influen-ceability (Eagly & Carli, 1981 ; Eagly & Wood, 1985), and aggression (Tieger, 1980 ; Hyde, 1984 ; Eagly & Steffen, 1986).

Maccoby and Jacklin listed "fairly well established differen-ces", "unfounded beliefs" and findings which in 1974 they still considered as "open questions" (judging their scientific credibility as insufficient or uncertain). We will question the notion of "well established" by taking a close look at the methodological and statistical decision-making rules used for judging and assessing the findings studied. In doing so, the notion of sex typicality should become a little clearer.

A Backdrop to the "Well Established"

Research in the field of sex differences can be characterized by both its wide range of application (all kinds of behaviour have been studied) and its internal fragility (revealed by the fact that every time someone tries to summarize or review previous research, many inconsistencies are discovered, and findings are shown to be unreliable and unstable).

As "well established" phenomena have grown out of the above environment, how can they remain unchallenged ? Block (1976b) spoke of their "erosion". It is obvious that, before 1966, the agreed-upon differences were greater in number than they are today. As they are subject to much disagreement (even to violent controversy) "well established" sex differences cannot be viewed as a finished product. Should they be totally rejected because of this ? Under what conditions can we consider something as knowledge (granted temporary), but real knowledge ?

The validity that can be attributed to the "well established" depends for the most part on how thoroughly the authors reviewing the phenomenon examined the conditions under which the data were produced, the methodological rules they applied, and the relevance of the theoretical basis on which the data base was selected.

Shields (1975) showed us that the study of sex differences is a scientific object that is eminently linked to the point of time in history when it took place, depending on the scientific environment and social circumstances of the times. The conditions under which it emerges partly determine the current theoretical framework and the research practices that model the results being evaluated. An analysis of current practices and their effect on findings is now well along its way (see for example Maccoby & Jacklin, 1974 ; Sherif, 1979 ; Grady, 1981 ; Hurtig & Pichevin, 1985). We might mention, however, that research biases have been pointed out at all levels, ranging from field observation to experimental setting, from data gathering to data interpretation, from diffusion to transmission of findings. Such biases should be taken into account, or even incorporated as modulators, in order to determine how much credit can be given to the conclusions drawn on this basis. Their generality is at stake.

We will deal with other strictly methodological difficulties. It appears that the major source of difficulty is the disparity of the available data. Actually, it is the diversity of the research (subfields of study, theoretical framework, methodology, measuring techniques) and the diversity of the dependent variables that make the data appear so incongruent. A researcher attempting to summarize must therefore define to what extent the studies, each generally totally independent, may be grouped together and compared for the purpose of combining the respective findings. Choosing a meaningful way of organizing and grouping them requires categorization of the research objects. This is what was called "the problem of conceptual rubrics" by Block (1976b), who writes "In regard to such large constructs as, for example, 'achievement', 'emphasis', 'impulsivity', 'self-esteem', 'conformity' or 'compliance', how shall we decide which studies properly relate to which concepts ?" (Block, 1984).

As stated by Maccoby and Jacklin (op. cit.), much of the research now used as a data base was not designed to study sex differences. The "incidental" nature of the information it has nevertheless provided, although it may be viewed as a screen against research biases specific to the study of sex differences themselves, is a serious handicap to the legitimacy of grouping together the findings. But incidental or not, we have to "make the best of" this data base, whose diversity reflects the variety of measuring conditions under which the findings were obtained-cf. for example Caplan et al., (1985) concerning the measurement of visual-spatial ability . This means that to be able to decide whether to group together or compare some of the findings, we must necessarily begin by examining what is being measured in each case. This single endeavour, in fact, raises two questions : what behavioural indices were chosen, and how was the study delimited ?

a. Behavioural indices. Discrepancies in the assessment of the reliability and sensitivity of behavioural indices, i.e. of their ability "to be supportable as an indicator of the underlying construct supposedly being evaluated" (Block, 1984) are one of the important sources of disagreement between authors (for instance, between Maccoby & Jacklin and Block). Two other points of disagreement are what behavioural unit to use (should we look for differences at the "molecular" level, act by act, or at the level at which these acts are organized into more complex behaviour patterns ?), and the problem of the degree of acceptable generalization (under what conditions can a difference observed on a specific index be transferred to a wider class of behaviour ?). Moreover, Block (1983) stressed that the differences only make sense when the relationships between different behaviours or spheres of behaviour, and their relationships to the different variables concerned, are accounted for.

b. The definition of what to study. We will illustrate this "definitional dilemma" (Caplan & al., 1985) with the example of research on aggression. To say, as Maccoby & Jacklin (op. cit.) did, "that males do appear to be the more aggressive sex, not just under a restricted set of conditions, but in a wide variety of behavioural indices" only holds up because they define aggression as "intent of individual to hurt another", a general enough definition to incorporate the diverse data considered. This is what Tieger (1980) calls their "linking definition of aggression across the various dependent measures". But by attempting, at all costs, to incorporate data, we may be losing some information that would contribute to our understanding. Macaulay (1985) suggested that this may be true of her analysis of the range of behaviour classified under the general label "aggression", whose diverse forms should be accounted for theoretically. "At one extreme the category includes physical battering, intentionally inflicted severe deprivation, vitriolic verbal abuse, and psychological torture, e.g. violence. At the other extreme are such things as unwelcome touching, the delivery of electric shocks in psychologists' laboratories and minor deprivations". Yet only the distinction between physical aggression and verbal aggression was considered from a theoretical point of view.

Along with Caplan & al. (op. cit.), we would like to conclude here that "in any area of research, if a construct has not been convincingly shown to be legitimate, it is premature and misleading to talk about group differences in the alleged construct".

The Significance of the Differences : Evaluation Criteria

Let us now put aside these disagreements, and take a look at the methodological and statistical criteria used to assert that there is a difference between the sexes in the areas analyzed.

We shall examine these criteria with regard to the differences considered by Maccoby and Jacklin to be "well established". They deal with cognitive abilities - verbal, visual-spatial, mathematical - and aggressive social behaviour.

Verbal ability. Girls are said to have a higher verbal ability than boys. They are said to learn to talk younger, and to be better at producing and comprehending language, especially during and after adolescence.

Visual-spatial ability. Boys are said to excel in such abilities from adolescence, although on certain types of tasks (measures of mental rotation and horizontality- verticality tests) their superiority is said to appear even before adolescence.

Mathematical ability. Here again, boys are said to be superior. According to recent data, the difference between boys and girls is more pronounced in algebra, but much less marked in arithmetic and geometry (Becker, 1983, cited by Deaux, 1985). These differences do, however, appear to vary significantly across populations.

Aggression. Boys are said to be more physically and verbally aggressive. Certain authors now agree that this difference appears quite early (before the age of six) (Maccoby & Jacklin, 1980 ; Hyde, 1984). Boys are also said to be aggressed against more often than girls.

Following the publication of The Psychology of Sex Differences, these differences have been the subject of intense controversy. Moreover, since 1974 many studies have shown that the differences in social behaviour are greater in number than estimated by Maccoby and Jacklin (Deaux, 1984). For example, in situations of social influence, women are said to conform to a greater degree than men to group pressure such as that used in the Ash and Sherif paradigm.

Statistical Significance and Frequency

A result is called "positive" if the difference between the two sex categories is statistically significant, which obviously implies a common ground for comparison. The direction of such a difference and its frequency of occurrence in a research corpus are two determining factors for Maccoby and Jacklin. Indeed, according to these authors, a sex difference that "holds up" must, above all, be relatively stable under identical or supposedly "equivalent" conditions. This does not mean, however, that the sex differences themselves are stable across time, since a given sex difference may only appear in a certain period in life, or under certain specific conditions. Maccoby and Jacklin's evaluation procedure is based on what other authors have called their "voting method" (Eagly & Carli, 1981). This method consists of counting the number of significant findings favouring each sex category and the number of findings indicating no difference. Differences whose significance level is less than or equal to .05 are retained as significant, and those for which it falls between .05 and .10 are considered as trends. As for the studies that do not show any differences, interpretation is more difficult. Although Maccoby and Jacklin were aware of the fact the "no amount of negative evidence proves that no difference exists", stating that remarks such as "no sex difference has been shown" is not equivalent to "there is no sex difference", they did not discard such findings from their account, including them in their summary tables under a column headed "none". This classification structures the data base thus constituted and alters the scope of the conclusions drawn from it. Let us illustrate this with an example. Among the 131 findings from studies on verbal ability conducted in test situations using a variety of standardized stimulus materials, only 28 % indicate that girls are significantly superior. That is not very impressive. However, if we analyze the data in another way by only considering the significant differences (38 % of the total findings), 74 % of them state that girls are superior. This attributes quite a different weight to the conclusions drawn. Different ways of reading the data, and the choices they involve, refer in fact to one and the same underlying methodological frame-work, which stresses the null hypothesis, an approach that some authors contest (for example, Block, 1976b), saying that the significant/non-significant dichotomy generates "hopeless inconsistencies". The statistical significance of the outcomes stating differences is in fact linked to the size of the samples used. But much of the experimental data analyzed by Maccoby and Jacklin was gathered with a relatively small number of subjects, making satisfactory significance levels difficult to attain. We are thus faced with a large number of findings that do not reach significance and thus cannot be relied upon, are uninterpretable, and lead to a loss of information.

To avoid this problem, and in order to obtain more reliable and objective results, analyses based on statistical techniques such as meta-analysis (Cooper & Rosenthal, 1980) have been developed. These methods make it possible to process an entire set of findings, not from "impressions gleaned by the reviewer from a reading of related studies" (Cooper, 1979) - an approach qualified by the author as "literary" - but by numerically combining the results of independent experiments that have in common either the same central hypothesis or some independent or dependent variables. In this way, a single set of numbers describing the body of related studies can be obtained, and, from this corpus one can establish the statistical probability that a given factor will have no effect, and evaluate the size of the effect. This method is advantageous in that it enables one to show effects which may otherwise be non-significant when considered in isolation, and to reveal interactions between variables studied in various independent studies. However, one must proceed with caution when employing meta-analysis procedures. In striving to be objective, these procedures may camouflage problems such as the pertinence and the legitimacy of regrouping studies. Moreover, there is disagreement as to the qualitative appreciation of equal numerical values and hence as to the importance of an effect (Hyde, 1984).

Among the estimations of effect size provided by these measures, the "d" index tells us how far apart the means of two groups are in terms of their common standard deviations. It is defined as the ratio of the difference between group means to the standard deviation of a group (assuming the standard deviations of the two groups are equal), or to the average of the standard deviations of the two samples, male and female.

Let us now analyze this way of assessing differences. It provides us with a means of comparing the magnitude of intergroup variation, represented by the distance between the means of each distribution, to the range of inter-individual variations within the population as a whole.

Magnitude of the Differences and Sex Effect Size

Maccoby and Jacklin did not underestimate the importance of this criterion. In some cases, for each of the studies considered they measured the magnitude of the difference between the mean values for males and females, providing a standard score : "the difference between the means divided by the weighted mean of the standard deviations of the two sex distributions". Plomin and Foch (1981) used the above numerical data to determine a weighted average standard score for a given group of studies. For example, for the 26 studies on verbal ability, the value of this index is less than 1/5 of the standard deviation of the whole population. That is not very much. Hyde (1981), analyzing the same data and some additional data in a meta-analysis, obtained very similar results to Plomin and Foch's : "the median value of 'd' is .24, that is the means of males and females were about one fourth of a standard deviation apart". Thus, no matter what measurement index is used, and regardless of what data are being analyzed, the difference between the mean scores on verbal ability is very small. Analyses of data on visual-spatial ability, mathematical ability, and aggressive behaviour lead us to the same conclusion, since the largest difference obtained is approximately 1/2 the standard deviation (Plomin & Foch, 1981 ; Hyde, 1981, 1984).

Thus, the best "well established" differences, which we could have expected to be great, are in fact only small differences between means. What about the other alleged differences, one might ask ? Cooper (1979) reanalyzed the findings on conformity cited by Maccoby and Jacklin. While Maccoby and Jacklin had concluded that these findings were inconsistent, classifying the alleged greater conformity of females in the "unfounded beliefs" category, Cooper's meta-analysis conducted on 47 studies grouped by situation, showed that the mean reactions of men and women to majority pressures differed from each other by about 1/4 of the standard deviation. This is neither more nor less than that found for some of the "well established" differences.

If the means show little difference, and grouping by sex only slightly reduces the range of interindividual variation, then sex does not contribute much as a subject variable to the observed variance. Analyses attempting to measure the size of the sex effect have shown that sex accounts for 1 to 5 % of the total variation, the "maxima" being attained for visual-spatial ability and aggressive behaviour. For mathematical ability, it accounts for either 1 % or 4 % of the variation, depending on the author. Hyde (1984) concluded that if these effects can legitimately be attributed to sex, the remaining 95 % (at least) of the variation must be due either to intra-sex group differences or to measurement errors ! If we accept that the measurements are correct, then we must admit that the within-sex variability is larger than the between-sex variability. It is known that intra-sex variability is great. Several researchers have compared the size of the difference between the responses given by two individuals chosen at random to that of two individuals of the same sex, whether two males or two females. These differences were very close.

Given that the overall effect is so weak, we would expect most studies with small samples not to reveal any differences. Indeed, Plomin and Foch showed that the smaller the number of subjects used in the experiments studied, the less likely one is to find overall significant differences between the two sex categories. The small differences established thus mainly describe large populations reduced to a mean score. Used as a variable to describe individuals, sex has little predictive value, so that if we only know a person's sex, we are not very likely to correctly predict his/her performance level.

Analysis of the distributions of the two sex categories indicates substantial overlap. For example, on verbal ability, the overlap is 90 % according to Plomin and Foch (quoting Cooper, 1977), and is between 81 % and 88 % on influence-ability according to Eagly and Carli (1981). In other words, knowing how an individual performs does not really increase our chances of correctly predicting his/her sex either.

Shape of the Distributions

As intergroup similarities and intragroup differences are

great, one final hypothesis remains to be considered. Are the small differences between the means due to extreme opposed values of the distributions, i.e. to those subgroups of individuals in either or both sexes that by differing strongly from their own sex category, differ significantly from the opposite sex ? The existence of such subgroups would in this case be the inter-sex differentiation element. This argument was put forth by Maccoby and Jacklin in their answer to Tieger (1980) on the question of whether differences in aggressiveness appear early or not : "The differences could mean either that most boys are somewhat more aggressive than most girls or that there are more boys than girls who are highly aggressive, with the bulk of boys being in the same range as girls. Another way of putting this last possibility is that the range is greater among boys so that in the small group of highly aggressive children boys are overrepresented". Maccoby and Jacklin tended to agree with the latter interpretation. The shape of the distributions and the number of extreme cases in each sex category should thus be studied systematically, something which had not yet been investigated. Maccoby and Jacklin spoke of some findings which indicated the range of the aggressive reactions, showing that it was wider for boys than for girls because of overrepresentation of males among extremely aggressive children.

Since then, additional data has been obtained by means of statistical extrapolation. The method used is quite simple. Choose a reference point on the overall distribution ; this point may be the mean, or another point that represents the desired selection cut-off point (for example, 5 % or 10 %). Then calculate how many members of each sex are located above or below this point. Eagly and Carli (1981) found that 60 % of the male subjects were less influenceable than the average woman respondent. Hyde (1981), using a highly selective value for visual-spatial ability of (5 %), found twice as many males (7.35 %) as females (3.22 %) above the value. She concluded that even small average sex differences "can generate rather large differences in proportions of males and females above some high cut-off point that might be required for outstanding performance of some occupations".

What conclusions can be drawn from this brief overview of how behavioural differences are evaluated ?
- It would be better to stress the similarities between the sexes than the differences in accounting for the psychological profiles of populations.
- The differences between the sexes are differences between means, which are ultimately very small. This does not mean, however, that they need not be explained when they are systematic.
- Sex as a differential factor appears generally to have little predictive value.
In the study of personality, we arrive at analogous conclusions

when examining research on the dimensions of masculinity and femininity. Granted, from a statistical standpoint, men have higher average scores on masculinity measures, and women have higher scores on femininity measures. However, as Deaux (1984) remarked, "the differences (...) are generally small in magnitude, and the distributions show considerable overlap. Thus, (...) sex-of-subject effects are limited but not inconsequential". And yet, femininity and masculinity scales, which are elaborated to best differentiate the two sexes, ought to provide the best indexes of psychological typicality ! Researchers have had to abandon the idea that masculinity and femininity are "psychological antinomies" (Spence & Helmreich, 1978),that they are two negatively correlated poles of one and the same dimension, each corresponding to a set of traits specific to one of the sexes. Today, masculinity and femininity are viewed as independent dimensions, and in this perspective an individual can be both masculine and feminine. Measurement scales that comply with this new model (such as the Bem Sex Role Inventory, 1974) define a more complex typology, one that stresses breaking away from the masculine/feminine dichotomy and introduces the notion of psychological androgyny.[1]

FROM STATISTICAL NORMS TO SOCIAL NORMS

We may therefore conclude that there are no cognitive, behavioural, or personality characteristics that are the exclusive lot of only one sex in humans, contrary to what everyday psychology tends to affirm. On the other hand, the social world still provides us with a very dichotomized picture of the activities of men and women, in particular through the sexual division of work. Sociological analyses have been conducted on this phenomenon. Can psychology account for this heterogeneous collection of data ?

Sex Typicality or Sex Conformity ?

Since there are no psychological characteristics that pertain to one sex, the reader can understand why we cannot talk about sex-typical behaviour. At best, a response, a behaviour pattern, etc... can be more or less in compliance with that which is observed on the average for a sex category i.e. with a sex standard that has been established statistically. In this perspective, could we describe a behaviour as masculine or feminine only according to how close it comes to this standard ? But if the distributions overlap so much, this criterion seems insufficient. Indeed, let us consider a response located in an area common to both sexes, and thus is not very far from either of the standards. What could we say about it ? It could be described as undifferentiated, in which case the sex factor can be considered irrelevant, or as being appropriate (conforming to one or other of the sexes), depending on the sex of the person who produced it. This example shows that behaviour can be judged as conforming to a given sex, but not as typical of that sex. Now let us consider a response that is far removed from the standard. Here again, its location is not, in itself, sufficient for determining its conformity or non-conformity. The direction of the deviation from the standard and the sex membership of the "atypical" subject are required in order to qualify his/her response. In practice, only those deviations from the standard of the subject's sex category that approach the extreme of the other sex are considered to correspond to non-conformity. It thus appears that the differentiating extremes in each distribution serve as the definitional poles of the masculine and feminine more than the means themselves. This may explain why, for example, substantial behavioural deviations from the sex standard observed in clinical studies (like those on female subjects who had been subject to high quantities of androgenous hormones during foetal life) are spoken of as an "apparent quantitative shift in a masculine direction" (Feder, 1984). The heterotypical here, rather than being defined by how close it is to the standard of the opposite sex, is defined by the single fact that the deviation occurs in the direction of the opposite pole. A girl may be much less aggressive than the average girl and could thus be viewed as not conforming. In practice, however, only those girls who are more aggressive are said not to conform to their sex. Such girls are labelled as "tomboys" and their level of aggressivity is judged as heterotypical. Although the typical does not exist, the heterotypical, on the other hand, can be defined as a particular case of non-conformity. Conformity and non-conformity can be symmetrical notions in a system where the notion of a standard makes sense. This is not true for typicality and heterotypicality, however, which we stated in the beginning of this chapter to be necessarily complementary. Perhaps that is why researchers in psychology speak of cross-sex behaviour, without using a truly symmetrical notion to describe behaviour that conforms to the standard, upon which no commentary is considered necessary since it fits into the order of things.

It is the normative weight of these poles that can be viewed as being responsible for unwarranted, but not meaningless generalizations, like changing the interpretation of a finding such as "boys are more aggressive than girls" into "boys are aggressive and girls are not".

Understanding the Variability of the Differences

Are the differences between the means ultimately of little interest ? Do they fail to account for the differences between the sexes that are observed in the social world, differences which are much greater than revealed by laboratory studies ? These two questions are much more consequential than they may appear at first. They reformulate the questions of what evaluation criteria to use and what value to attribute to conclusions where the small magnitude of a few sex differences prevails.

We are in no sense questioning the validity of the conclusions themselves, but rather wish to reflect upon the gap between everyday psychological knowledge, beliefs, stereotypes, and scientifically established facts ; to examine the relationships between laboratory outcomes and findings obtained outside the laboratory ; and to attempt to understand why commonly observed differences such as the inequality of job distribution among men and women are so great when the psychological differences are so small. Do such small psychological differences between the sexes partially explain the fact that out of the 455 professions listed in France, 167 of them employ less than 10 % women, whereas only 20 employ less than 10 % men ? Do the small differences in visual-spatial and mathematical ability partially explain why only 7 % of France's engineers were women in 1985 ? Can the fact that 45 % of the women who have a job are employed in the so-called "women's" professions be explained by their psychological conformity or their conformity to their sex category (i.e. to their lesser inter-individual variability) ?[2]

Nor do we wish to question the intrinsic validity of the evaluation criteria. We hope rather to show, by placing priority on the overall statistical validity of the available findings, that researchers do not deal with one of the nonetheless essential aspects of these findings : their instability, variability and "inconsistencies". At best, this aspect is considered as weakening the overall picture. Treating this fact as a finding, i.e. searching for the factors of such inconsistency, should contribute to identifying the factors of variation of the differences themselves, and to determining the conditions under which sex conformity emerges.

The Inconstancy of the Differences, the Inconsistency of the Findings

Many studies have revealed the variability of sex differences. Their variability itself is part of the picture : variations by age (Maccoby & Jacklin, 1974 ; Block, 1976a), culture (Whiting & Whiting, 1975), sociocultural subgroups and social classes (reviewed in Unger, 1979). Explanations for all of these variations, which have contributed to the nature/nurture issue, have been proposed. Variations across populations with diverse social origins, like some sex differences themselves, have been attributed to "sex typing", i.e. to the effects of the processes of socialization, sex role learning, and conformization to the prescribed sex standards of a given social group. It thus appears normal that sex conformity varies as a function of these social norms. Granted, the variables studied are still individual variables ; the sex differences observed and their variations across populations are still determined by the subject's internal characteristics, but these characteristics are viewed as acquired, shaped and modelled. Such variability is not inconsistency since it is irrefutable that social factors can account for it. When instability of findings is obtained under conditions where these social variables are controlled (i.e. when variability is ascribable neither to sex typing nor to maturational factors) the results may fairly be described as "precarious". This type of variability has often been called "inconsistency", and some feel that it should be eliminated, while others feel that it should be processed and then integrated along with other information. Cooper (1979) proposed that the term "inconsistency" should only be employed when research suggests or shows significantly opposing outcomes. This is rarely the case, but is nevertheless worth our attention. For example, out of the 94 findings on aggression examined by Maccoby & Jacklin (op. cit.), 52 of them indicated that males were more aggressive, and 5 that females were more aggressive (3 on children, 2 on adults). Thus, females are not always less aggressive than males, but under what conditions ? What factors cause them to demonstrate a level of aggressiveness that would be classified as heterotypical based on the statistical norms defined by the majority of the findings ? According to Cooper (op. cit.), any inconsistency may, in fact, indicate a not-yet-discovered interaction between variables.

A more common case is the discrepancy in the findings between the magnitudes of the sex difference for a given behaviour or ability, a difference which is greater, lesser or non-existent, depending on the study. Hyde (1984) for instance, noted that the magnitude of the difference in aggression in the studies published between 1978 and 1981 was less than that found in the studies published between 1966 and 1973. She proposes two possible types of explanation : (1) changes in socialization practices and sex standards, i.e. the liberalization of sex roles (but are the girls becoming more or the boys less aggressive ?), and (2) the fact that journals are now less reluctant to publish "negative" findings, and that data gatherers (experimenters as well as observers) perceive things differently. The second type of explanation gives credit to the stand taken by those who criticize the biased and artefactual nature of the findings used to base conclusions, invalidating the latter at the same time. On the other hand, both explanations considered jointly could validate a model wherein both the observer and the observed are subject to the same social sex norms, which evolve, the social norm being in one sense a parasite of the statistical norm.

Other discrepancies noted by many authors deal with the differences between field observations and experimental laboratory data, and the differences between different measurement techniques. For example, according to Hyde (1984), sex differences in aggression are "larger for naturalistic correlational studies than they are for experimental studies. They are larger when measured by direct observation, projective methods, or peer reports, and smaller when measured by self reports, parent, or teacher reports".

The Meaning of the Discrepancies

What do these discrepancies mean ? How do they affect the conclusions drawn ? Since the methodology applied affects findings so much, should these findings simply be classified as "pseudo phenomena" ? Are we right in thinking that "laboratory studies will produce more 'real' or 'better' effects than field studies (or vice versa for that matter)" (Unger, 1981) ? The laboratory situation has been thought to minimize the effect of sex norms (Eagly, 1978, cited by Unger, 1981) by creating a "social objectivity" norm, and by setting up a kind of sexual asepsis wherein the subjects somehow lose their sex. This could explain the near disappearance of the differences usually observed in everyday contexts. From an opposing perspective, field studies have been considered less reliable due to incomplete control of the possibly interfering variables involved. Rather than choosing between one of these two drawbacks, we suggest that context effects be considered as predictable by a model which defines sex as a social variable that "impinges upon the experimenter-subject relationship as well as upon relationships between subjects or between subjects and the social environment. An understanding of how such relationships are affected will enable us to predict more accurately when sex-related effects will appear in a study and when they will not" (Unger, 1981). Deaux & Major (1987) recently proposed a model that attempts to incorporate context effects. Along with these authors, we hypothesize that sex conformity, like the emergence of heterotypical behaviour, is dependent upon the characteristics of the social context. It is essentially a response to a situation immediately perceived as being subject to a norm, to a partner interaction situation, to the perception of a task, etc..., elements which could be considered regulators.

The Sources of Variability of the Findings

There are only a few systematic studies whose set objective was to determine the factors responsible for variability of findings in sex difference research. This lack is in keeping with the tendency to consider them as a consequence of interfering factors or research biases. We will examine the results of two meta-analyses, one on aggressive behaviour (Eagly & Steffen, 1986), and the other on "influenceability" (Eagly & Carli, 1981). These aimed to measure the effects of the characteristics on each of the studies included on the outcomes, i.e. on the magnitude of the difference and/or the size of the sex effect. Thus, each outcome is related to certain characteristics of a given study e.g. whether the author is a man or a woman, the journal publishing the data, etc... and refers to behaviour or responses produced in a given situation (laboratory or field, anonymous or not, with or without an observer, with a required response or free choice, involving social interaction with one or more male or female partners, conducted by a male or female experimenter, requiring free or imposed response forms, etc.). In addition to the above information used to classify the studies, the authors of these meta-analyses included indications on how other subjects unrelated to the studies and research situations perceive some of the aspects of those situations. They used questionnaires to gather data on (1) how members of each of the sexes perceive aggression and more precisely, the consequences of an aggressive act, and (2) just how interested members of each sex are in the topics approached in situations of social influence, i.e. in the sex typing of content inductions (for example, is a football game considered a masculine, feminine or neutral topic ?).

Context Variables

One context variable whose effect has often been considered is the researcher's sex (Unger, 1979). According to Eagly & Carli (1981), 79 % of the authors of "influenceability" research are men, who obtain sex differences indicating that women conform more, whereas in studies conducted by women, there are no sex differences. Authors may therefore produce a picture that favours their own sex.

Situation Variables

Certain characteristics of the situation make it possible to predict the size of the sex-of-subject effect. Thus, the sex typing of the topics chosen to evaluate "influenceability" influence the sex differences found for that particular behaviour. Women are known to conform more on "masculine" items, and men to conform more on "feminine" items (Goldberg, 1974, 1975, cited by Eagly & Carli, 1981). Eagly and Carli (op. cit.) noted that findings indicating higher influenceability in women tend to deal with topics judged as masculine.

Studying the research on aggression, Eagly and Steffen (1986) showed that the difference between the sexes (M>F) increases :
- when the partner is a male
- in laboratory situations
- on physical aggression as opposed to psychological aggression
- in semi-private contexts (i.e. in situations where the only witnesses are the partner-victim and the experimenter)
- when the aggression is required by the experimenter rather that chosen freely
- when the aggression is a reaction to strong rather than minimal provocation.

These variables interact strongly. For example, the partner's sex, whose effect varies across studies, becomes a good predictor of the sex-of-subject effect (i.e. whether a sex difference will be found and what will be its direction when associated with the other five variables listed above). These variables thus have a joint effect - multivariate models accounted for approximately 40 % of the variability in the available findings - even though the different experimental or field observation situations involved social contexts of limited variability. It is therefore quite understandable that in complex social contexts, non-conforming behaviour is found that can be accounted for by situation variables.

Furthermore, men and women responded quite differently on the questionnaires concerning their perception of consequences of an aggressive act. Women say they have more guilt feelings and experience much more anxiety with regard to the consequences of aggressive acts, judging them as more serious than men. These differences in judgment may be an additional source of variation in the magnitude of the sex difference. For example, Frodi & al. (1977) showed that in cases where aggressive behaviour is perceived as justified and prosocial, women can be just as aggressive as men. This suggests that aggression is regulated cognitively by its expected and evaluated consequences, in ways that mediate situational variables.

We can now predict that a field or laboratory situation in which all the variables take on the values that favour a difference is highly likely to produce the sex difference generally said to result only from the sex-of-subject effect. If we consider (as Eagly & Steffen, op. cit.) that the experimental paradigm most often used associates the conditions that account for the strongest sex-of-subject and sex-of-partner effects (namely, a laboratory situation involving physical aggression in a semi-private context with required aggressive action) it is easy to understand why a reversal of the sex difference is not likely to occur. The statistical norm in this case is clearly a function of the social characteristics of the situation.

Many studies on very young children have shown that such situation variables can at an early age cancel or even reverse generally admitted sex differences (see for example Jacklin & Maccoby, 1978 ; Serbin & al., 1979 ; Lloyd, 1987). In social life, however, everyday situations more often tend to reinforce differences between the sexes than to reverse them.

Sex-Related Effects

All this does not mean that the observed differences are caused by artefacts, even if they are relative and limited to certain contexts. The error that must not be made is attributing them only to the sex-of-subject factor when in fact they pertain to a network of effects that are related to sex, which plays the role of stimulus both to researchers and subjects (Hurtig & Pichevin, 1986). These effects are mediated by social interaction dynamics and by the cognitive processes of perception, evaluation, induction and attribution, all indissociable from the categorization process they presuppose and the accompanying stereotyping processes (Ashmore & Del Boca, 1986).

For example, the presence or absence of observers, whether real or assumed, or the situation's public/private dimension, which makes the sex identity of the subjects more or less salient, leads to responses that do or do not conform (to a greater or lesser degree) to the stereotypes. Thus, women may tend to conform more when a male expert is present during experimentation (Eagly, 1978, cited by Unger, 1981). Under anonymous conditions, individuals of both sexes may produce the opposite sex's response patterns. Kidder & al. (1977) showed that in reward allocation situations, men as well as women comply with traditional norms - and with the statistical norms of past literature on the subject - when they know their decisions will be made public, whereas they adopt the opposite sex's behaviour when they are sure their response will remain anonymous. The authors attribute this reversal to the subject's being "relieved of the burdens of femininity and masculinity".

While the occurrence of heterotypical behaviour was "symmetrical" for both sexes in the above study, it is not always so. Frodi & al. (1977) found that the anonymity factor also had an effect on whether girls adopted certain aggressive, heterotypical behaviour : "There is some evidence that the condition of anonymity (or deindividuation) in a situation may have the same effect as that of justification for aggression". In this case, this factor acts in the same direction in both sexes, increasing aggressiveness in boys and girls.

It appears that in daily living, sex conformity pressure, although strong for both sexes, is stronger for males, and that the adoption of heterotypical behaviour is more frequent in females. There are more social sanctions towards males who act outside of their sex role and who produce behaviour or manifest traits considered characteristic of females (Maccoby & Jacklin, 1974). This can be explained by the fact that the behaviour and traits attributed to females are valued less socially, as shown by the correlations of masculine and feminine items with social value and desirability (Spence & Helmreich, 1978). The adoption of cross-sex behaviour may thus be more profitable to females. Several analyses conducted on psychological androgyny seem to confirm this phenomenon. Indeed, androgyny (as an individual psychological characteristic) appears to have little weight compared to the effects of a social context dominated by masculine values (Locksley & Colten, 1979). The degree of masculinity is thus a better predictor than androgyny of both men's and women's self-esteem, adjustment and psychological health (Taylor & Hall, 1982).

We can see that immediate context effects may be the expression of more general effects linked to social structure, here the social hierarchy and the power relationships between the sexes. The cognitive processes of perception, evaluation, induction, and attribution carried out in a given situation are in part prede-termined by the structural characteristics of the sex typing that prescribes the sex roles, namely gender attributes (Sherif, 1982).

CONCLUSION

It therefore appears difficult (in light of available research findings) to speak of sex-typical behaviour, and more fitting to speak of sex-conforming behaviour. We can probably speak of gender-typical behaviour. The term gender is to be understood here in all of its diverse utilisations and thus includes :
- attributes that are culturally perceived as appropriate to males or females
- traits through which the sex label acts as a trigger of differential behaviour
- that which each individual attributes to or expects of other people depending on their sex label
- that which an individual attributes to him/herself depending on his/her own sex category.

This definition outlines new research perspectives (see Deaux, 1985) only touched upon in this conclusion. We might emphasize at this point that this definition leads us to consider sex - which is both a social and a cognitive category - as a socio-cognitive label. The socio-cognitive variables that must be considered and isolated in this perspective are those of sex information processing. As such they are cognitive variables, but their role in cognitive functioning is highly dependent upon interpersonal and intergroup interaction processes and upon sociocultural contexts. By taking these variables into account, reinterpretation of the available findings is possible, giving them new coherence. Their role in and importance to the active construction of reality may indeed explain why gender-typical behaviour can be defined as soon as the first act which identifies and assigns to one or the other of the sex categories is performed (see e.g. Kessler & McKenna, 1978 ; Grady, 1979 ; Fagot & al., 1986).

It would therefore be false to believe that, at one extreme, there are stereotypes that delineate two impervious worlds, one masculine, one feminine, and at the other, there is one world of real behaviour, without differentiation by sex. Indeed, if typical behaviours, which are expected, are indeed perceived, they may also be produced. Such expectations may concern oneself (thus orienting the subject's own behaviour) or they may concern others and operate as so-called "self-fulfilling prophecies" (Zanna & Pack, 1975). Finally, the interplay of reciprocal expectations in situations of social interaction may reinforce or prevent the emergence of gender-typical behaviour.

NOTES

1. Psychological androgyny is used to describe men and women with approximately equal proportions of both masculine and feminine psychological characteristics. The notion of androgyny, which has sometimes been proposed as an alternative to emphasizing sex differences, and even as a prescription for social change, has often been highly criticized both from a methodological and theoretical standpoint (Pedhazur & Tetenbaum, 1979 ; Lubinski & al., 1983).
2. These data are taken from a report entitled "Femmes en chiffres", published in 1987 by the I.N.S.E.E.

REFERENCES

Ashmore, R.D. and Del Boca, F.K. (Eds.) (1986). The social psychology of female-male relations : A critical analysis of central concepts. Orlando, FL : Academic Press.

Bem, S.L. (1974). The measurement of psychological androgyny. Journal of Consulting and Clinical Psychology, 42, 155-162.

Bem, S.L. (1981). Gender schema theory : A cognitive account of sex typing. Psychological Review, 88, 354-364.

Block, J.H. (1976a). Debatable conclusions about sex differences. Contemporary Psychology, 21, 517-522.

Block, J.H. (1976b). Issues, problems, and pitfalls in assessing sex differences : A critical review of "The Psychology of Sex Differences". Merrill-Pallmer Quarterly, 22, 283-308.

Block, J.H. (1983). Differential premises arising from differential socialization of the sexes : Some conjectures. Child Development, 54, 1335-1354.

Block, J.H. (1984). Sex role identity and ego development. San Francisco, CA : Jossey-Bass Publishers.

Caplan, P.J., MacPherson, G.M. and Tobin, T. (1985). Do sex related differences in spatial abilities exist ? A multilevel critique with new data. American Psychologist, 40, 786-799.

CNIDF-INSEE (1987). Femmes en chiffres. Paris : Publications de l'INSEE.

Cooper, H.M. (1979). Statistically combining independent studies : A meta-analysis of sex differences in conformity research. Journal of Personality and Social Psychology, 37, 131-146.

Cooper, H.M. and Rosenthal, R. (1980). Statistical versus traditional procedures for summarizing research findings. Psychological Bulletin, 87, 442-449.

Deaux, K. (1984). From individual differences to social categories : Analysis of a decade's research on gender. American Psychologist, 39, 105-116.

Deaux, K. (1985). Sex and gender. Annual Review of Psychology, 36, 49-81.

Deaux, K. and Major, B. (1987). Putting gender into context : An interactive model of gender-related behavior. Psychological Review, 94, 369-389.

Eagly, A.H. and Carli, L.L. (1981). Sex of researchers and sex-typed communications as determinants of sex differences in influenceability : A meta-analysis of social influence studies. Psychological Bulletin, 90, 1-20.

Eagly, A.H. and Steffen, V.J. (1986). Gender and aggressive behavior : A meta-analytic review of the social psychological literature. Psychological Bulletin, 100, 309-330.

Eagly, A.H. and Wood, W. (1985). Gender and influenceability : Stereotype versus behavior. In O'Leary V.E., Unger R.K. & Wallston B.S.(Eds.), Women, gender, and social psychology (pp. 225-256). Hillsdale, NJ : Erlbaum.

Fagot, B.I., Leinbach, M.D., and Hagan, R. (1986). Gender labeling and the adoption of sex-typed behaviors. Developmental Psychology, 22, 440-443.

Fairweather, H. (1976). Sex differences in cognition. Cognition, 4, 231-280.

Feder, H.H. (1984). Hormones and sexual behavior. Annual Review of Psychology, 35, 165-200.

Frodi, A., Macaulay, J. and Thome, P.R. (1977). Are women always less aggressive than men ? A review of the experimental literature. Psychological Bulletin, 84, 634-660.

Grady, K.E. (1979). Androgyny reconsidered. In Williams J.H. (Ed.), Psychology of women : Selected readings (pp. 172-177). New York : Norton.

Grady, K.E. (1981). Sex bias in research design. Psychology of Women Quarterly, 5, 628-636.

Hall, G.S. (1904). Adolescence (Vol.2).New York : Appleton.

Hurtig, M.C. and Pichevin, M.F. (1985). La variable sexe en psychologie : donné ou construct ? Cahiers de Psychologie Cognitive, 5, 187-228.

Hurtig, M.C. and Pichevin, M.F. (1986). Conclusions. In Hurtig M.C. & Pichevin M.F. (Eds.), La différence des sexes. Questions de psychologie (pp. 321-331). Paris : Editions Tierce.

Hyde, J.S. (1981). How large are cognitive gender differences ? A meta-analysis using w2 and d. American Psychologist, 36, 892-901.

Hyde, J.S. (1984). How large are gender differences in aggression ? A developmental meta-analysis. Developmental Psychology, 20, 722-736.

Jacklin, C.N. and Maccoby, E.E. (1978). Social behavior at thirty-three months in same-sex and mixed-sex dyads. Child Development, 49, 557-569.

Kessler, S.J. and McKenna, W. (1978). Gender : An ethnomethodological approach. New York : Wiley.

Kidder, L.H., Bellettirie, G. and Cohn, E.S. (1977). Secret ambitions and public performances : The effects of anonymity on reward allocations made by men and women. Journal of Experimental Social Psychology, 13, 70-80.

Lloyd, B. (1987). Social representations of gender. In Bruner J. & Haste H. (Eds.), Making sense : The child's construction of the world (pp. 147-162). London : Methuen.

Locksley, A. and Colten, M.E. (1979). Psychological androgyny : A case of mistaken identity ? Journal of Personality and Social Psychology, 37, 1017-1031.

Lubinski, D., Tellegen, A. and Butcher, J.N. (1983). Masculinity, femininity and androgyny viewed and assessed as distinct concepts. Journal of Personality and Social Psychology, 44, 428-439.

Macaulay, J. (1985). Adding gender to aggression research : Incremental or revolutionary change ? In O'Leary V.E., Unger R.K. & Wallston B.S. (Eds.), Women, gender, and social psychology (pp. 191-224). Hillsdale, NJ : Erlbaum.

Maccoby, E.E., and Jacklin, C.N. (1974). The psychology of sex differences. Stanford, CA : Stanford University Press.

Maccoby, E.E. and Jacklin, C.N. (1980). Sex differences in aggression : A rejoinder and reprise. Child Development, 51, 964-980.

Morgan, M.J. (1980). Influences of sex on variation in human brain asymmetry. Behavioral and Brain Sciences, 3, 244-245.

Pedhazur, E.J. and Tetenbaum, T.J. (1979). Bem Sex Role Inventory : A theoretical and methodological critique. Journal of Personality and Social Psychology, 37, 996-1016.

Plomin, R. and Foch, T.T. (1981). Sex differences and individual differences. Child Development, 52, 383-385.

Serbin, L., Connor, J.M., Burchardt, C.J. and Citron, C.C. (1979). Effects of peer presence on sex-typing of children's play behaviors. Journal of Experimental Child Psychology, 27, 303-309.

Sherif, C.W. (1979). Bias in psychology. In : Sherman J.A. & Beck E.T. (Eds.), The prism of sex (pp. 93-133). Madison, WI : The University of Wisconsin Press.

Sherif, C.W. (1982). Needed concepts in the study of gender identity. Psychology of Women Quarterly, 6, 375-398.

Shields, S.A. (1975). Functionalism, Darwinism, and the psychology of women : A study of a social myth. American Psychologist, 30, 739-754.

Spence, J.T. and Helmreich, R.L. (1978). Masculinity and femininity : Their psychological dimensions, correlates and antecedents. Austin, TX : University of Texas Press.

Taylor, M.C. and Hall, J.A. (1982). Psychological androgyny : Theories, methods, and conclusions. Psychological Bulletin, 92, 347-366.

Tieger, T. (1980). On the biological basis of sex differences in aggression. Child Development, 51, 943-963.

Unger, R.K. (1979). Female and male : Psychological perspectives. New York : Harper & Row.

Unger, R.K. (1981). Sex as a social reality : Field and laboratory research. Psychology of Women Quarterly, 5, 645-653.

Whiting, B.B. and Whiting, J.W.M. (1975). <u>Children of six cultures.</u> Cambridge, MA : Harvard University Press.

Zanna, M.P. and Pack, S.J. (1975). On the self-fulfilling nature of apparent sex differences in behavior. <u>Journal of Experimental and Social Psychology,</u> <u>11</u>, 583-591.

3 Prenatal Gonadal Influences on Mouse Sociosexual Behaviours

Frederick S. Vom Saal *

A basic assumption in research on sexual differentiation has been that after differentiation of the gonads, testosterone (or its intracellular metabolites : 17β-oestradiol and 5α-dihydrotestosterone) mediates the process of masculinization (the induction of species-specific masculine characteristics). Active defeminization (the loss of species-specific feminine characteristics) is proposed to occur in male rodents (but not primates, see e.g. Karsch & al., 1973) as a result of intracellular conversion of testosterone to oestradiol within specific areas of the brain. Examples of defeminization of specific traits in response to testosterone secretion by the male testes during early life in mice and rats are : the loss of the capacity to respond to elevated oestradiol by exhibiting a surge in LH or the sexually-receptive posture (lordosis) when mounted by a stud male (Gorski, 1979). A second hormone secreted by the foetal testes, Mullerian inhibiting hormone, is also required for active degeneration (defeminization) of the derivatives of the Mullerian ducts (fallopian tubes, uterus, and upper part of the vagina ; Jost, 1972).

In contrast to the active processes of masculinization and defeminization in males, the process of feminization in females was presumed to be passive. Circulating gonadal steroids were thus presumed to play little role in the development of the female phenotype. The findings presented below provide evidence contradicting some of these hypotheses. Specifically, the hypothesis that gonadal steroids have little effect on the ontogeny of the "normal" female phenotype is not supported by our findings. The concept that only circulating androgen influences sexual differentiation will also be challenged.

* Division of Biological Sciences and Department of Psychology, University of Missouri-Colombia, Missouri 65211, USA.

In this chapter, masculinization and feminization will refer to development of a different suite of characteristics in males and females that render them capable of surviving, competing for mates, and successfully producing and rearing young (see Figure 1). Some examples of traits which differ between males and females are : specific morphological traits (external and internal genitalia ; Wilson & al., 1981), physiological traits (energy allocation strategies and utilization of nutrients ; Perry & al., 1979 ; Wade & Gray, 1979 ; Perrigo & Bronson, 1985), and hypothalamic, liver, kidney and genital tissue enzyme systems (De Moor & al., 1973 ; Einarsson & al., 1973 ; Siiteri & Wilson, 1974 ; Naftolin & al., 1976 ; Lamartiniere, 1979 ; Bardin & Catterall, 1981). Specific behaviours include competing for territories, attracting mates, mating, and the care and defense of offspring (Vom Saal, 1981 ; 1984). There are also many other sex differences in behaviour, such as lower activity levels in males than females (Perrigo & Bronson, 1985 ; Kinsley & al., 1986) and differential learning abilities (Beatty, 1979).

Different conclusions are arrived at when a comprehensive approach to the study of the ontogeny of all aspects of the masculine vs. feminine phenotype is taken as opposed to basing one's model on studies of the ontogeny of just one component of these complex processes, such as copulatory behaviour. A decrease

PHENOTYPE

FEMININE	MASCULINE
Gonads	
oogenesis ovulation steroid secretion	spermatogenesis steroid secretion
Accessory Reproductive Organs	
fallopian tubes uterus vagina mammary glands	epididymus seminal vesicle prostate penis
Reproductive Behaviour	
proceptivity receptivity Infanticide/parenting interfemale aggression postpartum aggression	mounting/intromitting ejaculating Infanticide/parenting intermale aggression

Figure 1 : Selected characteristics of organisms, labelled as either masculine or feminine, which influence whether an individual can compete for mates and produce and raise healthy young.

in the capacity to exhibit one set of behaviours (for example, mounting, intromitting and ejaculating in males or lordosis in females) cannot always be viewed from the simplistic perspective that a generalized demasculinization or defeminization has occurred. A particular hormonal milieu might result in a decrease in copulatory behaviour while producing an opposite effect on some other index of masculinization or feminization, such as intrasex aggression, with the result that the likelihood of a male or female achieving dominance is increased although copulatory behaviour is decreased. In this situation the likelihood of successfully producing and rearing young might actually be highest in animals with the lowest rates of sexual behaviour. Experiments will be described in which numerous complex relationships between different components of both the masculine and feminine phenotype have been identified in rats and mice.

SOURCES OF TESTOSTERONE IN RODENT FOETUS

Rodents are polytocous mammals. Around the end of the second week of gestation in rats, mice and hamsters, the testes in male embryos differentiate and begin secreting testosterone (Block & al., 1971 ; Pontis & al., 1979 ; 1980 ; Feldman & Bloch, 1978 ; Vomachka & Lisk, 1986). Organization of the ovaries in females occurs later in foetal life (shortly before parturition ; reviewed in : Vom Saal & Finch, 1988). The testes are not the only source of testosterone during foetal life in rodents. The placenta in primates metabolizes androgen to oestrogen due to the presence of aromatase (Siiteri & Thompson, 1975). In contrast, placentae in rats and mice secrete androstenedione (a weak androgen) and testosterone (Jackson & Albrecht, 1985 ; Soares & Talamantes, 1982 ; 1983 ; Vreeburg & al., 1983) rather than oestrogen (Sybulski, 1969). The importance of this finding is that both male and female foetuses are exposed to supplemental androgen of placental origin. However, the activity of enzymes involved in the synthesis of testosterone (17β-hydroxylase and C17,20-lyase) is significantly higher in placentae collected from female than from male foetuses on Day 18 of pregnancy in CF-1 mice (Vom Saal & al., 1987).

We also find that female foetuses have fairly high titres of circulating testosterone during the last four days of pregnancy in mice, although male foetuses have about 2.5 times higher circulating levels of testosterone than do female foetuses and pregnant females (Figure 2 ; unpublished observation). Similar sex differences in testosterone exposure during foetal life have been reported in rats (Weisz & Ward, 1980), hamsters (Vomachka & Lisk, 1986), monkeys (Resko, 1975) and humans (Reyes & al., 1974).

Individual differences in the circulating levels of testosterone in both male and female mouse foetuses (and, most likely, also in rat foetuses) are correlated with the sex of the foetuses next to the animal being examined. Thus, male foetuses have higher blood

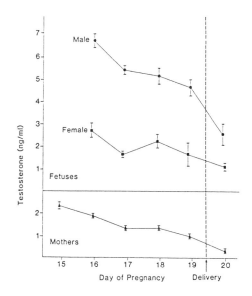

<u>Figure 2</u> : The serum concentrations of testosterone (±SEM) measured by RIA in pregnant female CF-1 mice and male and female foetuses during the latter part of pregnancy and the day of birth (unpublished observation).

levels of testosterone than do females, and animals positioned <u>in utero</u> between male foetuses (2M males or 2M females) have significantly higher blood titres of testosterone than do animals of the same sex positioned between female foetuses (OM males and OM females) ; animals situated between a male and a female (1M) represent 50 % of the population of males and females and are intermediate in their blood testosterone levels. Exactly the opposite relationship is observed for blood levels of oestradiol, with females having significantly higher levels than males, and OM animals having significantly higher levels than 2M animals of the same sex (Figure 3 and 4 ; Vom Saal & al., 1988). During foetal life in humans, females also have higher blood titres of oestradiol than do males (Reyes & al., 1974).

There are thus three major sources of testosterone in foetal mice : males have a source of testosterone, the testes, that result in higher circulating concentrations of testosterone in males relative to females. Testosterone is also secreted by the placentae, but the placenta of a female may secrete higher titres of testosterone than that of a male (Vom Saal & al., 1987). Finally, both males and females may receive supplemental testosterone as a result of having implanted in the uterus next to one or two male foetuses (1M and 2M animals).

Since female mouse foetuses develop in the presence of high circulating levels of testosterone yet remain fertile, it is of

Intrauterine Position

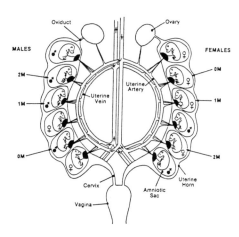

Figure 3 : Schematic diagram of the uterine horns and uterine
loop arteries and veins of a pregnant mouse. The labels : OM, 1M,
and 2M refer to the number of male foetuses an individual is next
to (2M = between 2 males, 1M = between a male and a female,
and OM = between 2 females). Arrows within the vessels indicate
the direction of blood flow.

Day 18 of Pregnancy

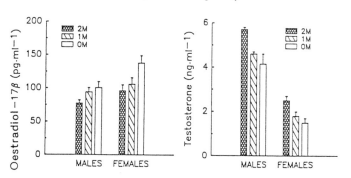

Figure 4 : The serum concentrations of testosterone and oestradiol
in OM, 1M and 2M male and female CF-1 mouse foetuses (mating
= Day O). For both testosterone and oestradiol, males vs. females,
and within each sex, intrauterine position differences were
statistically significant (P<.O5 ; Vom Saal & al., 1988).

interest to determine the possible basis for the absence of characteristics typical of males. Three possibilities are : (1) the levels of testosterone, while high, may still be below threshold values, (2) activity of androgen receptors may be lower in females than in males, and (3) development of some tissues requires conversion of testosterone to more potent metabolites (for example, oestradiol in the sexually dimorphic nucleus of the preoptic area (SDN-POA ; Dohler & al., 1984 ; 1986) and dihydrotestosterone in the external genitalia (George & Wilson, 1988). Females have lower levels of activity of the enzymes that metabolize testosterone to these more potent hormones than do males during sexual differentiation ; specifically, hypothalamic aromatase activity (Naftolin & al., 1976 ; MacLusky & Naftolin, 1981) and perineal tissue 5α-reductase activity (Kelch & al., 1971 ; Siiteri & Wilson, 1974). During the differentiation of these tissues, even in the face of exposure to equal circulating titres of testosterone, females would thus not respond to the same degree as would males.

INDIVIDUAL DIFFERENCES IN PHENOTYPE IN FEMALE RATS AND MICE DUE TO INTRAUTERINE POSITION

The intrauterine position phenomenon provides a unique method of correlating testosterone and oestradiol exposure during foetal life with postnatal traits in rats and mice (we also have found differences due to intrauterine position in pigs ; Rohde-Parfet & al., 1988). Embryos of each sex implant randomly in terms of their intrauterine position (Vom Saal, 1981), so there should be no systematic difference in genotype between animals from different intrauterine positions. The relationship of intrauterine position to postnatal characteristics in rats and mice will be reviewed. In these experiments females are time mated, and young are delivered by Cesarean section shortly before normal parturition. The intrauterine position of each foetus is identified by a toe-clipping procedure, and litters are fostered to mothers that had delivered normally within the preceding 24 hr.

Differential Exposure to Testosterone in Female Foetuses as a Model for Studying Masculinization

There are two basic strategies that have been used to study the postnatal consequences of having developed in different intrauterine positions. One approach has involved using females from different intrauterine positions as models for studying the ontogeny of male behaviours. This strategy involves treating females with testosterone in adulthood to induce heterotypical behaviours (those characteristic of the opposite sex, such as attack toward a male or mounting of a sexually-receptive female). An assumption underlying these experiments has been that effects due to differential exposure to testosterone during foetal life

would be latent in females ; these effects would only be observed
if the neural areas that had been influenced by testosterone during
early life were subsequently "activated" by exposure to exogenous
testosterone at the time of testing for a specific behaviour. This
strategy is useful for learning about the basis of differences in
sensitivity to the activational effects of steroids.

An example of the above strategy involved treatment of
pregnant rats and mice with antiandrogens (flutamide and
cyproterone acetate), which may act by inhibiting binding of
androgens to androgen receptors. The perineal tissue separating
the anus and genital papilla becomes the scrotum in males and
serves as a bioassay for the levels of testosterone to which
animals are exposed during foetal life. In both rats (Clemens &
al., 1978) and mice (Vom Saal & Bronson, 1978) the length of the
space separating the anus and genital papilla is longer in 2M than
in 0M females at birth (1M females are intermediate between 0M
and 2M females). Exposure to an antiandrogen reduced the mean
length of the perineal tissue separating the anus and genital
papilla at birth in female as well as male offspring, and all males
and females were indistinguishable (Clemens & al., 1978). In mice,
this was due to anogenital distance at birth being reduced in 2M
and 1M females, such that these females were similar to 0M
females on this measure (Vom Saal, 1976).

Prenatal antiandrogen treatment also resulted in the female
offspring exhibiting a decreased response to testosterone in
adulthood : flutamide-treated female rats exhibited less masculine
sexual behaviour in response to adult treatment with testosterone
than did females not previously exposed to flutamide (Clemens &
al., 1978). In addition, cyproterone-exposed female mice (2M, 1M
and 0M) were relatively insensitive to the activating effects of
adult treatment with testosterone in terms of attack toward a
male intruder (intermale-like aggression), similar to normal 0M
females (Vom Saal, 1976). Taken together, these findings reveal
that it is possible to detect an effect of inhibiting the action of
testosterone in female foetuses. On the other hand, these
experiments have not addressed whether there is any consequence
of prenatal exposure to an antiandrogen in terms of altering some
female-typical behaviour.

Differences in the Timing of Puberty, Oestrous Cycles, Attrac-
tiveness, Sexual Behaviour, and Pheromonal Cues due to
Intrauterine Position : Relevance for Reproductive Success

The second strategy for examining the consequences of
developing in different intrauterine positions in mice and rats has
involved designing experiments with regard to the potential
significance to reproductive success of individuals in natural
habitats. This approach differs from the first strategy in that
pharmacological manipulations during perinatal life are not
employed. I have used both strategies for examining the

intrauterine position phenomenon. However, the differences due to intrauterine position that have been identified without any pharmacological manipulations have led to different conclusions concerning the role of sex steroids in the development of the "normal" male and female phenotype than have pharmacological studies.

Mice are extremely successful opportunists that have, as a species, occupied a wide variety of niches with many different types of social structures. Phenomenal differences in population size are found in studies of Mus musculus domesticus, with densities ranging from a few mice per hectare in fields to thousands per 100 cubic metres in corn ricks (Southwick, 1958 ; Christian, 1971). Weanling mice often disperse from their natal environment, but some mice remain and compete for resources within the natal deme. It is possible that the intrauterine position phenomenon provides phenotypic variation such that there are animals within each litter with characteristics that render them likely to be successful at each of these strategies, thus maximizing the fitness of the mother. This leads to the hypothesis that the intrauterine position phenomenon evolved because it is adaptive to have animals within a litter vary in phenotype due to a random developmental event, regardless of the degree of underlying genetic variation. Even in highly inbred strains of mice, such as the C57BL/6J strain, there is substantial unexplained variability in numerous traits, some of which might be accounted for by the intrauterine position phenomenon (Felicio & al., 1984 ; Svare & al., 1984 ; Vom Saal & Finch, 1988).

Many of the following comparisons were made using only OM and 2M females, since in all studies in which 1M females have been tested, they have been intermediate between OM and 2M females in their characteristics. The basic approach in all of the following experiments was to compare OM and 2M females in terms of characteristics that could influence their reproductive success in a natural environment.

In comparisons of the attractiveness of OM and 2M female mice to males, most (82 %) males entered a chamber containing a OM female rather than a 2M female, OM females were mounted more than 2M females (Vom Saal & Bronson, 1978 ; 1980a), and most males inseminated OM females prior to inseminating 2M females (Rines & Vom Saal, 1985). OM female mice had higher lordosis quotients (number of lordoses/number of mounts), an index of sexual receptivity, than did 2M females (Figure 5 ; Rines & Vom Saal, 1985) ; similar findings were obtained in comparisons of OM and 2M female rats (unpublished observation). First oestrus (puberty) occurred significantly earlier in OM female rats than in 2M female rats (Vom Saal, 1981). Finally, in both mice and rats, OM females had shorter oestrous cycles (mostly 4 days in length) than did 2M females (5-7 days in length ; Vom Saal & Bronson, 1980b ; Vom Saal, 1981). In terms of sexual receptivity, the emission of cues that attract males, the timing of puberty, and length of oestrous cycles, elevated blood testosterone (and/or

lower oestradiol) levels during foetal life appears to have defeminized 2M females (see Figure 6).

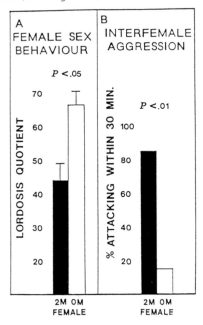

Figure 5 : A. Lordosis quotient (lordoses/mounts x 100) for adult
OM and 2M female mice that were ovariectomized and injected
with oestradiol benzoate 48 hr prior to testing and progesterone
4 hr prior to being placed with a stud male (Rines & Vom Saal,
1985). B. The percent of OM and 2M females (paired when in
dioestrus and matched for age and weight) that attacked and
established dominance over the opponent (based on 20 pairs that
fought ; Vom Saal & Bronson, 1978).

All of the above comparisons of OM and 2M females suggest
that developing between male foetuses places 2M females at a
reproductive disadvantage relative to OM females, but this may
not be the case in some environments. Ovulation in female mice is
regulated by pheromones secreted into the urine of both males and
females. 2M females are less responsive than OM females to the
effects of pheromones produced by other females that inhibit
ovulation (Vom Saal, 1981 ; Vom Saal & al., 1981 ; unpublished
observation). In an environment with a high density of female
mice, therefore, 2M females enter puberty (ovulate and mate) at a
younger age than do OM females. In addition, the postpubertal
oestrous cycles of OM females are longer than those of 2M
females. However, exactly the opposite occurs if females are
housed near or with a male, but without other females around :

2M females enter puberty later and have longer postpubertal oestrous cycles than do OM females (Vom Saal, 1981 ; Vom Saal & Bronson, 1978 ; 1980b ; Vom Saal & al., 1981).

An important aspect of the above findings is that adolescent OM female mice only have short cycles if the density of females in the environment is low. The advantage conferred on adolescent OM females of having short oestrous cycles, and a higher likeli- hood of ovulating, thus only operates when population density is low. Something that is often not appreciated is that predation rates of mice in the wild are typically very high. In terms of population dynamics, variables that influence the likelihood of ovulating and mating at the earliest possible age (early adolescence) have the greatest impact on reproductive success in natural populations. Differences between OM and 2M females that are only revealed at advanced ages (such as differences in the decline in reproductive capacity with age (Rines & Vom Saal, 1985 ; Vom Saal & Moyer, 1985) are likely to have little impact on population dynamics, although they are important to models of aging (Vom Saal & Finch, 1988).

Morphology			
• anogenital separation	OM females	shorter than	2M females
Physiology			
• age at puberty	OM females	younger than	2M females
• oestrous cycle	OM females	shorter than	2M females
• cessation of reproduction	OM females	older than	2M females
Stimulus Characteristics			
• attract males	OM females	greater than	2M females
• arouse males			
Sensivity to Oestrogen			
• sexual receptivity	OM females	greater than	2M females
Sensitivity to Testosterone			
• mounting	OM females	less than	2M females
Behaviour			
• aggression toward females	OM females	less than	2M females
• postpartum aggression			

Figure 6 : Summary of some of the comparisons of OM and 2M female mice.

Interfemale Aggressive Behaviour and Reproductive Success

The studies described above lead to the prediction that when population density is low, OM females have a reproductive advan- tage over 2M females due to accelerated puberty, shorter oestrous cycles, greater attractiveness to males, and higher levels of sexual receptivity when mounted. However, an intriguing possibility is that exposure to elevated titres of testosterone during foetal life by 2M females might actually result in behavioural characteristics (together with insensitivity to density-dependent pheromonal cues) that render them the most likely to successfully produce and raise offspring in high density populations. In high density populations of mice, interfemale competition for nest areas and aggressiveness in defense of nursing young may be important to reproductive success (Lloyd & Christian, 1969), and 2M females are more aggressive than OM females.

When adult 2M and OM female CF-1 mice in dioestrus were paired, 2M females were more aggressive, and most 2M females (85 %) became dominant over the OM female opponent (Figure 5). The level of aggression between non-lactating female mice is not as intense as that observed between males of most mouse stocks, but biting, chasing and clear submissive behaviour by the defeated female were observed (Vom Saal & Bronson, 1978). In an experiment examining postpartum (maternal) aggression, 2M female mice were more aggressive toward an intruder than OM females (Vom Saal & Bronson, 1978). These findings have been causally related to differences in prenatal exposure to testosterone by the experiment of Mann and Svare (1983) : pregnant female mice were treated with a low dose of testos-terone so that normal development of the internal and external genitalia occurred in female offspring. After mating and delivering young when adults, the prenatally testosterone-exposed females were more aggressive toward an intruder than control females in defense of their young.

In wild mice both males and females kill young (in laboratory stocks of mice, females typically do not kill pups), although whether this occurs depends on both reproductive state and dominance status (Vom Saal & Howard, 1982 ; Vom Saal, 1985 ; McCarthy & Vom Saal, 1985, 1986a ;b ; McCarthy & al., 1986 ; Vom Saal, 1988). Aggression by lactating female mice toward other females, as well as males, is thus a critical behaviour to ensure survival of the litter due to the threat posed by both males and other females (Parmigiani & al., 1988a ; b).

Hormonal Basis of Aggression by Females

Elevated titres of testosterone during foetal life lead to an elevation of aggressiveness in 2M female mice. 2M female mice were found to be more aggressive than OM females while in a variety of hormonal states : (1) when OM and 2M females were paired in dioestrus while cycling normally, (2) when a female intruder was placed into the home cage of a OM or 2M female after delivering young (postpartum aggression), and (3) when treated with testosterone and examined for aggression against a male opponent (Vom Saal & Bronson, 1978 ; Rines & Vom Saal, 1985 ; reviewed in Vom Saal, 1983a). 2M females thus are more aggressive than OM females regardless of hormonal state at the time of testing ; activation by one specific hormone does not appear to be required to observe aggression. When a behaviour is infuenced by hormones during early life but does not require activation by a specific hormone to be observed, the behaviour is referred to as having been "organized" during early life (Vom Saal, 1983a).

A major conclusion from comparisons of female mice from different intrauterine positions is that some aspects of behaviour that have traditionally been used as markers of masculinization, such as aggression toward other adults, occur naturally in both sexes. However, the characteristics of the behaviour (target site attacked : head vs. flanks) and both stimulus and hormonal control, differ dramatically between males and females (Vom Saal, 1983a ; Parmigiani & al., 1988 a ; b). 2M females thus cannot be labelled as masculinized because they are more aggressive than OM females.

The finding that female mice do fight with each other and also attack males in some situations is actually not an uncommon observation, although until recently aggression between females was simply dismissed as unimportant. In some studies of freely-growing populations of mice, only a few aggressive females produced and weaned young successfully (Retzlaff, 1938 ; Lloyd & Christian, 1969 ; reviewed in Vom Saal, 1983a ; Vom Saal, 1984). A unique aspect of the model that androgens also play a role in "normal" female sexual differentiation is the proposition that feminization is a complex active process. For example, a trait such as intrasex aggression, which appears to be increased by perinatal exposure to elevated titres of androgen in both male and female mice (but has only been viewed as an index of masculinization), may actually influence the likelihood of both males and females mating and successfully rearing young.

Interfemale Aggression and Reproductive Success in Seminatural and Natural Environments

Based on experiments in which mice were allowed to breed freely in confined areas and then all but one male was removed at various times, Christian and co-workers have concluded that, at least in some environments, females are highly aggressive and territorial. Interestingly, when all but one male was removed, both the aggressive, territorial females and non-aggressive females mated and produced litters. Unlike less aggressive females, however, the territorial females continued to patrol their territories and guard against intruders rather than care for their young, which resulted in the death of their young (Yasukawa & al. 1985 ; Chovnick & al., 1987).

An experiment that indirectly addressed the relevance of the intrauterine position phenomenon for reproductive success of females in a natural environment was recently reported by Ims (1987). When weanling voles (Clethrionomys rufocanus) were placed onto an island from which all voles had been cleared, most females that survived and successfully produced young came from litters containing a high proportion of females. The conclusion was that these females would have most likely been OM females. Ims proposed that for the voles in this study, reproductive success may have been higher for females that were not aggressive, similar to the findings of Chovnick & al. (1987) with house mice described above. The impact of elevated aggressiveness on reproductive success in female rodents is thus likely to be a function of population density and social structure.

INDIVIDUAL DIFFERENCES IN PHENOTYPE IN MALE RATS AND MICE DUE TO INTRAUTERINE POSITION

Intrauterine position of male mice is correlated with both plasma and amniotic fluid levels of oestradiol and testosterone (Figure 4 ; Vom Saal & al., 1983, 1988). We have examined the accessory reproductive organs as well as sexual behaviour, intermale aggression, and behaviour toward young in OM and 2M male mice. Both sexual behaviour and the volume of the sexually dimorphic nucleus of the preoptic area (SDN-POA) have been examined in OM and 2M male rats.

2M males had initially been predicted to show the highest rates of male sexual behaviour (mounting, intromitting and ejaculating) when paired with a sexually-receptive female, since 2M males have the highest levels of testosterone during foetal life. However, the results were opposite to this prediction : OM male mice exhibited more mounts and intromissions than 2M males when paired with a sexually-receptive female (Figure 7 ; Vom Saal & al., 1983). Similarly, when OM and 2M male rats were tested to sexual satiety (30 min without a mount when paired with a sexually-receptive female), OM males had higher scores on all measures of sexual behaviour ; in particular, OM males exhibited more ejaculations than 2M males. These same OM male rats had a significantly larger volume of the SDN-POA than did the 2M males (unpublished observation).

Gonadally intact, adult OM and 2M male mice were examined for prostate weights, and OM males had significantly heavier prostates than did 2M males (unpublished observation). The interesting aspect about all of the above findings that will be discussed below is that OM male mice have lower levels of testosterone but higher circulating levels of oestradiol than do 2M males during foetal life (Figure 4). Given the similarity of the findings concerning the correlation between intrauterine position and adult sexual behaviour in both mice and rats, it seems likely that the differences in foetal testosterone and oestradiol levels due to intrauterine position in mice also occur in rats.

In another comparison OM and 2M male mice that had been castrated at birth (to eliminate any effect of postnatal exposure to gonadal steroids) were treated with oestradiol and progesterone and placed with a stud male. Most OM males elicited mounting by the stud and exhibited lordosis when they were mounted ; too few

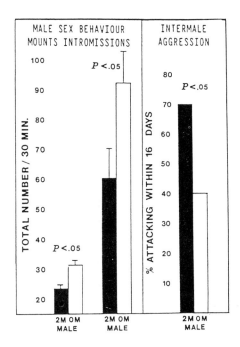

Figure 7 : Sexual behaviour (number of mounts and intromissions during a 30-min test) in adult, gonadally-intact CF-1 OM and 2M male mice when paired with a sexually-receptive female. Intermale aggression exhibited by adult OM and 2M male mice toward a male intruder within 16 days of being implanted with a silastic capsule containing testosterone (tests for aggression occurred every other day) ; OM and 2M males were gonadecto-mized at Cesarean delivery (Vom Saal & al., 1983).

2M males were mounted to assess their lordosis behaviour (Vom Saal & al., 1983). We have also observed that OM females (which are exposed to higher circulating oestradiol and lower circulating testosterone than 2M females during foetal life) exhibit higher lordosis indices than 2M females in both rats (unpublished observation) and mice (Figure 5 ; Rines & Vom Saal, 1985). Thus, a correlate of exposure to elevated oestradiol and low levels of testosterone in OM male mouse foetuses is the facilitation of sexual behaviour in adulthood : either male sexual behaviour when exposed to testosterone or female sexual behaviour when exposed to oestrogen and progesterone.

The response to treatment with testosterone in OM and 2M male mice castrated at birth has been assessed in terms of the duration of treatment required to induce intermale aggression. 2M males were more responsive to testosterone than were OM males :

more 2M males exhibited aggression within 16 days of testosterone treatment (Figure 7 ; Vom Saal & al., 1983). The behaviour of adult, gonadally-intact OM, 1M and 2M male mice toward newborn young also revealed marked differences (Figure 8). Most 2M males exhibited parental behaviour (built a nest, retrieved the pups to the nest, groomed the pups, and kept them warm by hovering over them). In contrast, most OM males exhibited infanticide (the killing of preweanling young ; Vom Saal, 1983b, 1988).

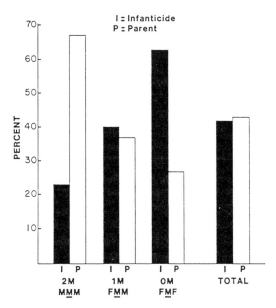

Figure 8 : The proportion of adult OM, 1M and 2M male CF-1 mice exhibiting infanticide (I) or parental behaviour (P) in response to the presence of a newborn pup placed in the home cage (Vom Saal, 1983b).

While 2M males are more responsive to testosterone in terms of the induction of aggression toward other adult males, they are the least likely to be aggressive toward young. This finding reveals that constructs such as aggression are not unitary phenomena. Numerous experimental studies in which testosterone exposure during early life has been manipulated, in addition to comparisons of OM and 2M males, have revealed that elevation of testosterone during early life leads to an enhancement of adult intermale aggression but reduces the likelihood of animals exhibiting aggression toward young (Vom Saal, 1983b, 1984, 1988 ; Svare & al., 1984). Exactly what role, if any, variation in oestradiol titres during foetal life plays in mediating the differences in intermale aggression and behaviour toward young between OM and 2M males is, as yet, unclear (Vom Saal, 1988).

In other studies, OM and 2M males were castrated and implanted with silastic capsules containing equal concentrations of testosterone or dihydrotestosterone (gonadally intact animals were also examined). 2M males had heavier seminal vesicles (blotted wet weight) and preputial glands than OM males when treated with testosterone, but not when treated with dihydrotestosterone. Treatment with dihydrotestosterone significantly increased the weight of seminal vesicles in OM males but had the same effect as testosterone in 2M males. This finding suggested that the lower weight of the seminal vesicles in OM males when gonadally intact or treated with testosterone was due to a deficit in the activity of 5α-reductase, which metabolizes testosterone to dihydrotestosterone (testosterone acts as a prohormone for dihydrotestosterone synthesis in the seminal vesicles). Confirming this hypothesis was the finding that gonadally intact, as well as gonadectomized and testosterone-treated, 2M males had significantly higher seminal vesicle 5α-reductase activity than did OM males (unpublished observation).

One mechanism via which differences in foetal hormone exposure can modulate adult morphology and behaviour is thus to influence the activity of the intracellular enzymes that amplify the effect of testosterone by metabolizing it to the more potent hormones : oestradiol and dihydrotestosterone. The basis of the difference between castrated OM and 2M males in the activation of intermale aggression after treatment with testosterone remains to be elucidated, but differences in 5α-reductase activity or androgen receptor function are possible.

Circulating Oestrogen, Plasma Binding Proteins, and Intracellular Aromatization during Foetal Life

One issue raised by our findings with OM and 2M males is whether circulating oestradiol has an effect on the development of the prostate, sexual behaviour, and SDN-POA volume during foetal life. If circulating testosterone mediated differences between OM and 2M males after intracellular aromatization in these tissues (Dohler & al., 1984 ; Marts & al., 1987), then a 2M male would have a larger prostate, SDN-POA volume, and higher rate of sexual behaviour than a OM male. However, if circulating oestradiol can enter specific cells in the prostate, neurons mediating male sexual behaviour and neurons in the SDN-POA, then the elevated oestradiol in the circulation of OM males could interact with testosterone within these cells and enhance "masculinization" of these tissues in OM males relative to 2M males. The ratio of testosterone to oestradiol in the circulation might thus be important, and in both male and female mouse foetuses, this ratio varies significantly as a function of intrauterine position (Figure 9).

A central tenet of the above hypothesis is that one cannot

Day 18 of Pregnancy

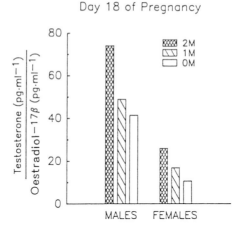

Figure 9 : The ratio of testosterone to oestradiol in serum collected from OM, 1M and 2M male and female CF-1 mouse foetuses
on Day 18 of pregnancy. Based on the data in Figure 4.

assess masculinization based only on the titres of testosterone to
which males are exposed during foetal life. This is certainly
contrary to what would typically be expected. This hypothesis
is likely to apply only to tissues in which testosterone serves as a
prohormone for aromatization to oestradiol. Once inside a target
cell, oestradiol binds to oestrogen receptors associated with the
nucleus (Welshons & Gorski, 1986). Thus, masculinization of tissues
containing aromatase would presumably involve activation of
intracellular oestrogen receptors rather than androgen receptors.
    A basic assumption of the aromatization hypothesis is that
oestradiol must be derived from the intracellular metabolism of
testosterone. Testosterone is thus proposed to be a prohormone in
these tissues. Circulating oestradiol is bound to the plasma
protein, alphafetoprotein (AFP), an alpha-globulin that binds
oestrogen but not androgen or other steroids (MacLusky &
Naftolin, 1981 ; Westphal, 1986). Thus, circulating oestradiol is
presumed to be inhibited from entering cells and "interfering" with
normal development of the female phenotype in female foetuses
(for example, in OM female foetuses, which have the highest
circulating levels of oestradiol). Steroids bound to plasma proteins
(other than albumin) are generally deemed unable to enter cells
(Pardridge, 1981). An interesting aspect of this model is that it
appears that AFP binds oestrogen only in rodents, but not other
species that have been examined (MacLusky & Naftolin, 1981).
This model thus does not explain the absence of effects of circulating oestrogens in species other than rodents.
    Most mammals have one plasma protein, sex steroid binding
globulin (SSBG), which binds both oestradiol and testosterone.

although the capacity for these steroids to pass from the blood into cells does differ (oestradiol dissociates from SSBG and thus enters cells more rapidly than does testosterone ; Pardridge, 1981). In most mammals, therefore, both oestradiol and testosterone can enter cells and potentially influence the course of differentiation (depending on the presence of specific intracellular receptors). In rodents, however, the available evidence is that there is no SSBG (Corvol & Bardin, 1973 ; Renoir & al., 1980 ; Bardin & al., 1981 ; Siiteri & al., 1982 ; Petra & al., 1983). In rodents, testosterone should only be weakly bound to albumin and freely enter cells, while oestradiol should be tigtly bound to AFP and thus be unable to enter cells (Pardridge, 1981).

Some aspects of the aromatization hypothesis have been challenged. Toran-Allerand (1984) has argued that oestradiol bound to AFP can enter certain target tissues ; AFP is found in some neural areas while mRNA for AFP is only found in the liver. The hypothesis is that there are receptors for AFP on specific neurons that allow AFP to be transported into these cells via receptor-mediated endocytosis (Vom Saal, 1983c). Even in rodents, circulating oestrogen could thus influence the differentiation of selected tissues (those able to internalize AFP). One aspect of this hypothesis is the prediction that within a target cell, oestradiol would dissociate from AFP and bind to oestrogen receptors (which have a higher affinity for oestradiol than does AFP).

The conclusion from comparisons of the foetal hormone titres, adult sexual behaviour, and SDN-POA volume in OM and 2M males is that circulating oestradiol can enter selected neurons (and possibly other tissues such as the prostate) and influence differentiation of these tissues during early life (Figure 10). The elevation in circulating oestradiol in OM male foetuses (due to being positioned between female foetuses) thus presumably enhances the "masculinization" of these tissues, even though background levels of circulating testosterone are low in OM males relative to 2M males. One possibility is that the intracellular levels of aromatase may be low enough so that the 25 % difference between OM and 2M male foetuses in circulating testosterone is damped out in terms of the amount of oestradiol produced via aromatization. However, the high-affinity oestrogen receptors in these neurons would presumably render them responsive to small changes in circulating oestradiol, if the oestradiol is actually able to enter cells in proportion to the concentration in the blood (a question that is still unresolved). This model is testable, and experiments are underway to examine this hypothesis.

The similarity of the seminal vesicle and onset of intermale aggression findings (2M>OM males) in contrast to prostate weights, SDN-POA volume and masculine sexual behaviour (OM>2M males) suggests a common mechanism underlying each set of responses. Intermale aggression and seminal vesicle weight are positively correlated with elevated testosterone levels during

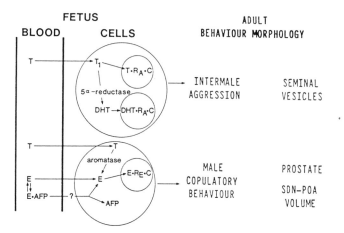

Figure 10 : Schematic diagram depicting the possible mechanisms by which testosterone and oestradiol in the blood influence cell differentiation during foetal life. This model predicts that two consequences of elevated blood testosterone in 2M males during foetal life are increased intermale aggression and seminal vesicle weight in adulthood. Consequences of elevated blood oestradiol in OM males (either free or bound to alpha-fetoprotein, AFP) during foetal life are increased male copulatory behaviour, prostate weight, and sexually dimorphic nucleus of the preoptic area (SDN-POA) volume in adulthood.

foetal life, and 2M males show a greater responsivity to testosterone than do OM males on both measures. These findings lead to the hypothesis that testosterone acts without being aromatized to oestradiol in neural areas mediating intermale aggression (Figure 10). We find no evidence for the presence of oestrogen receptors in seminal vesicles of CF-1 mice, and differences between OM and 2M males in seminal vesicle weight can be accounted for by differences in 5α-reductase activity (unpublished observation).

Both the low circulating concentrations of oestradiol (relative to androgen) during foetal life and the low affinity of androgen receptors for oestradiol make it unlikely that circulating oestradiol influences development of tissues with androgen receptors. In contrast, even small changes in circulating oestradiol might be detected in tissues with high-affinity oestrogen receptors, such as the prostate (uterine-like oestrogen receptors have been identified in prostates of adult CF-1 male mice, and adult OM male mice have heavier prostates than do 2M males).

The Aromatization Hypothesis

It has generally been accepted that circulating testosterone serves as the substrate for intracellular synthesis of oestradiol via aromatization in some neural areas, and that oestradiol both "masculinizes" and "defeminizes" these neural areas during perinatal life (MacLusky & Naftolin, 1981). This model is thus referred to as the "aromatization hypothesis". If aromatization is inhibited (via treatment with blockers of aromatase) or binding of oestradiol to oestrogen receptors is blocked (via treatment with oestrogen receptor blockers), then differentiation into the female phenotype is supposed to be observed. However, this hypothesis has only been thoroughly examined in rats in relation to the ontogeny of male and female sexual behaviour, the increase in the volume of the SDN-POA, and defeminization of the capacity to surge LH in response to oestrogen treatment (Gorski, 1979 ; MacLusky & Naftolin, 1981 ; Dohler & al., 1984 ; 1986). Whether this hypothesis holds for any other behaviour (such as intermale aggression and behaviour toward young) is controversial, since few studies examining whether oestradiol could exert organizational effects on these behaviours have been conducted, and they have typically involved the use of very high doses of oestrogen (Edwards & Herndon, 1970 ; Samuels & al., 1981).

The picture that emerges from studies addressing the aromatization hypothesis in which pharmacological manipulations have been employed is far from clear. An interesting finding is that treatment with an androgen receptor blocker, such as cyproterone acetate or flutamide, during early postnatal life blocks the development of male sexual behaviour (Clemens & al., 1978), as well as inhibits defeminization of the capacity to exhibit female sexual behaviour (Neumann & Elger, 1966 ; Gladue & Clemens, 1978). However, the aromatization hypothesis predicts that only oestrogen receptor blockers, not androgen receptor blockers, should inhibit masculinization and defeminization of sexual behaviour.

The ontogeny of male and female sexual behaviour and the LH-surge system is likely to involve a complex interaction within individual neurons of testosterone (or its reduced metabolite, dihydrotestosterone), which would bind to androgen receptors, and intracellular metabolism of testosterone to oestradiol, which would then bind to oestrogen receptors (Dohler & al., 1986). It is still unclear whether individual neurons involved in regulating sexual behaviour or the preovulatory surge in LH might have both androgen and oestrogen receptors. A number of cell lines, such as MCF-7 cells (a breast cancer-derived cell line) have both androgen and oestrogen receptors (Horwitz & al., 1975). In rat prostate, there are androgen and oestrogen receptors (Jung-Testas & al., 1981), aromatase (Marts & al., 1987) and 5α-reductase (Wilson & al., 1981). Both aromatase and 5α-reductase are also found in some neural areas, although the distribution of these enzymes in the brain differs considerably (Selmanoff & al., 1977). However, it is considered unlikely that any single cell would have 5α-reductase and aromatase activity, since 5α-reduced androgens inhibit aromatase activity (Siiteri & Thompson, 1975).

One hypothesis has been that steroids have the greatest effect on the process of masculinization of sexual behaviour in rats during prenatal life, while defeminization of the capacity to exhibit female sexual behaviour occurs primarily during the early postnatal period. Thus, treatment of male rats with the aromatization inhibitor ATD (1,4,6-androstatriene-3,17-dione) during early postnatal life blocked the defeminization of female sexual behaviour that would normally have occurred in intact males but did not reduce masculine sexual behaviours when animals were tested in adulthood (Davis & al., 1979). Administration of ATD to pregnant rats did interfere with masculinization of sexual behaviour in male offspring (Gladue & Clemens, 1980). However, this same prenatal treatment with ATD also interfered with defeminization of the lordosis response in both males and females (Clemens & Gladue, 1978).

Further complicating this issue is the report that postnatal inhibition of aromatization (Booth, 1978) or treatment with a synthetic oestrogen (RU 2858 ; Booth, 1977) interfered with both masculinization and defeminization of sexual behaviour in rats. The picture that emerges from these findings is that differences in methodology (such as in the studies by Booth, 1978 and Davis & al., 1979) can lead to different conclusions concerning the timing and hormonal regulation of masculinization and defeminization of sexual behaviour (Dohler & al., 1986). One additional finding that raises questions concerning the significance of all of the above studies of defeminization of sexual behaviour is the report of Södersten et al., (1983). Sex differences in the capacity for oestradiol to induce female sexual behaviour in rats that were gonadectomized in adulthood were observed only if oestradiol was administered tonically (via silastic capsules). No sex differences in lordosis were observed when oestradiol was administered in a pulsatile manner (via injection every 12 hr) prior to testing.

CONCLUSION

The process of differentiation of a masculine or feminine phenotype in mammals is far more complex than proposed in initial models, which stated that elevated testosterone levels lead to masculinization, while the absence (or low levels) of testosterone lead to the persistence of feminine traits. Data concerning prostate weight, SDN-POA volume, and rates of mounting, intromitting and ejaculating suggest that masculinization is enhanced in OM males. Findings concerning seminal vesicle weight and the onset of intermale aggression suggest exactly the opposite (Figure 11). Even when focusing just on the ontogeny of sexual behaviour, conclusions concerning the mechanisms by which testosterone results in the facilitation of

| INDICES | DEFEMINIZATION | |
	FEMINIZATION ←	MASCULINIZATION →
Morphology		
• seminal vesicle weight	OM Male	2M Male
• SDN-POA volume	2M Male	OM Male
Behaviour		
• aggression	OM Male	2M Male
• sexual activity	2M Male	OM Male

Figure 11 : Comparisons of OM and 2M male rats : sexual activity and volume of the sexually dimorphic nucleus of the preoptic area (SDN-POA ; unpublished observation) and OM and 2M male mice (sexual activity, intermale aggression, and seminal vesicle weight ; Vom Saal & al., 1983). The type of male listed under masculinization had the highest score for each measure.

male copulatory behaviours and the inhibition of female copulatory behaviours differ based on the methods used.

Taken together, these findings provide a clear example of why one cannot generalize concerning phenomena as complex as masculinization and defeminization based on studies of only one characteristic, such as sexual behaviour. These terms should only be used to refer to the ontogeny of all traits that distinguish males and females of a particular species. Perhaps the most important finding in studies of the intrauterine position phenomenon with both males and females is that all animals are able to reproduce successfully in an optimum laboratory environment. There is indirect evidence, however, from studies of rodents in both seminatural and natural environments (Chovnick & al., 1987 ; Ims, 1987) that differences, such as have been observed in experimental comparisons of OM, 1M and 2M males and females, can influence reproductive success.

ACKNOWLEDGEMENTS

Support for much of the research described in this chapter was provided by a grant from the National Science Foundation (DCB-8518094). I thank P. Franks and W. Welshons for their helpful comments during the preparation of this chapter.

REFERENCES

Bardin, C. and Catterall, J. (1981). Testosterone : A major determinant of extragenital sexual dimorphism. Science, 211, 1285-1294.

Bardin, C., Musto, N., Gunsalus, G., Kotite, N., Cheng, S.-L., Larrea, F. and Becker, R. (1981). Extracellular androgen binding proteins. Annual Review of Physiology, 43, 189-198.

Beatty, W. (1979). Gonadal hormones and sex differences in nonreproductive behaviors in rodents : Organizational and activational influences. Hormones and Behavior, 12, 112-163.

Block, E., Lew, M. and Klein, M. (1971). Studies on the inhibition of fetal androgen formation : testosterone synthesis by fetal and newborn mouse testes in vitro. Endocrinology, 88, 41-46.

Booth, J. (1977). Sexual behavior of neonatally castrated rats injected during infancy with oestrogen and dihydrotestosterone. Journal of Endocrinology, 72, 135-141.

Booth, J. (1978). Effects of the aromatization inhibitor androst-4-ene-3,6,17-trione on sexual differentiation induced by testosterone in the neonatlly castrated rat. Journal of Endocrinology, 79, 69-76.

Chovnick, A., Yasukawa, N. Monder, H. and Christian, J. (1987). Female behavior in populations of mice in the presence and absence of male hierarchy. Aggressive Behavior, 13, 367-375.

Christian, J. (1971). Population density and reproductive efficiency. Biology of Reproduction, 4, 248-294.

Clemens, L. and Gladue, B. (1978). Feminine sexual behavior in rats enhanced by prenatal inhibition of androgen aromatization. Hormones and Behavior, 11, 190-201.

Clemens, L., Gladue, B. and Coniglio, L. (1978). Prenatal endogenous androgenic influences on masculine sexual behavior and genital morphology in male and female rats. Hormones and Behavior, 10, 40-53.

Corvol, P. and Bardin, C. (1973). Species distribution of testosterone-binding globulin. Biology of Reproduction, 8, 277-282.

Davis, P., Chaptal, C. and McEwen, B. (1979). Independence of the differentiation of masculine and feminine sexual behavior in rats. Hormones and Behavior, 12, 12-19.

De Moor, P., Verhoeven, G. and Heyns, W. (1973). Permanent effects of foetal and neonatal testosterone secretion on steroid metabolism and binding. Differentiation, 1, 241-253.

Dohler, K., Hancke, J., Srivastava, S., Hofmann, C., Shrine, J., and Gorski, R. (1984). Participation of estrogen in female sexual differentiation of the brain : neuroanatomical and behavioral evidence. Progress in Brain Research, 61, 99-117.

Dohler, K., Coquelin, A., Davis, F., Hines, M., Shryne, J. and Gorski, R. (1986). Aromatization of testicular androgens in physiological concentrations does not defeminize sexual brain function. In : Systemic Hormones, Neurotransmitters and Brain Development, Dörner G., McCann S., Martini L. (Eds.), Karger, Basel, pp. 28-35.

Edwards, D. and Herndon, J. (1970). Neonatal estrogen stimulation and aggressive behavior in female mice. Physiology and Behavior, 5, 993-995.

Einarsson, K., Gustafsson, J. and Stanbag, A. (1973). Neonatal imprinting of liver microsomal hydroxylation and reduction of steroids. Journal of Biological Chemistry, 248, 4987-4997.

Feldman, S. and Bloch, E. (1978). Developmental pattern of testosterone synthesis by fetal rat testes in response to Luteinizing Hormone. Endocrinology, 102, 999-1007.

Felicio, L., Nelson, J. and Finch, C. (1984). Longitudinal studies of estrous cyclicity in aging C57BL/6J mice : II. Cessation of cyclicity and the duration of persistent vaginal cornification. Biology of Reproduction, 31, 446-453.

George, F. and Wilson, J. (1988). Sex determination and differentiation. In : Physiology of Reproduction, Knobil E. and Neill J. (Eds.), Raven Press, New York, pp. 3-26.

Gladue, B. and Clemens, L. (1978). Androgenic influences on feminine sexual behavior in male and female rats : defeminization blocked by prenatal antiandrogen treatment. Endocrinology, 103, 1702-1709.

Gladue, B. and Clemens, L. (1980). Masculinization diminished by disruption of prenatal estrogen biosynthesis in male rats. Physiology and Behavior, 25, 589-593.

Gorski, R. (1979). The neuroendocrinology of reproduction : an overview. Biology of Reproduction, 20, 111-127.

Horwitz, K., Costlow, M. and McGuire, W. (1975). MCF-7 : a human cancer cell line with estrogen, androgen, progesterone and glucocorticoid receptors. Steroids, 26, 785-795.

Ims, R. (1987). Determinants of competitive success in Clethrionomys rufocanus. Ecology, 68, 1812-1818.

Jackson, J. and Albrecht, E. (1985). The development of placental androstenedione and testosterone production and their utilization by the ovary for aromatization to estrogen during rat pregnancy. Biology of Reproduction, 33, 451-457.

Jost, A. (1972). A new look at the mechanisms controlling sex differentiation in mammals. Johns Hopkins Medical Journal, 130, 38-53.

Jung-Testas, I., Groyer, M., Bruner-Lorand, J. Hechter, O., Baulieu, E. and Robel, P. (1981). Androgen and estrogen receptors in rat ventral prostate epithelium and stroma. Endocrinology, 109, 1287-1289.

Karsch, F., Dierschke, D. and Knobil, E. (1973). Sexual differentiation of pituitary function : Apparent difference between primates and rodents. Science, 179, 484-486.

Kelch, R., Lindholm, U. and Jaffe, R. (1971). Testosterone metabolism in target tissues : 2. Human fetal and adult reproductive tissues, perineal skin and skeletal muscle. Journal of Clinical Endocrinology, 32, 449-456.

Kinsley, C., Miele, J., Konen, C., Ghiraldi, L. and Svare, B. (1986). Intrauterine contiguity influences regulatory activity in adult female and male mice. Hormones and Behavior, 20, 7-12.

Lamartiniere, C. (1979). Neonatal estrogen treatment alters sexual differentiation of hepatic histidase. Endocrinology, 105, 1031-1035.

Lloyd, J. and Christian, J. (1969). Reproductive activity of individual females in three experimental freely growing populations of house mice (Mus musculus). Journal of Mammalogy, 50, 49-59.

Mann, M. and Svare, B. (1983). Prenatal testosterone exposure elevates maternal aggression in mice. Physiology and Behavior, 30, 503-507.

MacLusky, N. and Naftolin, F. (1981). Sexual differentiation of the central nervous system. Science, 211, 1294-1303.

Marts, S., Padilla, G. and Petrow, V. (1987). Aromatase activity in microsomes from rat vental prostate and Dunning R3327H rat prostatic adenocarcinoma. Journal of Steroidal Biochemistry, 26, 25-29.

McCarthy, M. and Vom Saal, F. (1985). The influence of reproductive state on infanticide by wild female house mice. Physiology and Behavior, 35, 843-849.

McCarthy, M. and Vom Saal, F. (1986a). Inhibition of infanticide after mating in wild male house mice. Physiology and Behavior, 36, 203-209.

McCarthy, M. and Vom Saal, F. (1986b). Infanticide by virgin CF-1 and wild male house mice (Mus musculus) : Effects of age, prolonged isolation and testing procedure. Developmental Psychobiology, 19, 279-290.

McCarthy, M., Bare, J. and Vom Saal, F. (1986). Infanticide and parental behavior in wild female house mice (Mus musculus) : Effects of ovariectomy, adrenalectomy, and administration of oxytocin and prostaglandin. Physiology and Behavior, 36, 17-23.

Naftolin, F., Ryan, K. and Davies, J. (1976). Androgen aromatization by neuroendocrine tissues. In : Subcellular Mechanisms in Reproductive Neuroendocrinology, Naftolin F., Ryan K. and Davies J. (Eds.), Elsevier Pub., Amsterdam, pp. 347-355.

Neumann, F. and Elger, W. (1966). Permanent changes in gonadal function and sexual behaviour as a result of early feminization of male rats by treatment with an antiandrogenic steroid. Endokrinologie, 50, 209-225.

Pardridge, W. (1981). Transport of protein-bound hormones into tissues in vivo. Endocrine Review, 2, 103-123.

Parmigiani, S., Brain, P., Mainardi, D. and Brunoni, V. (1988a). Different patterns of biting attack generated when lactating female mice (Mus domesticus) encounter male and female conspecific intruders. Journal of Comparative Psychology, 102, 287-293.

Parmigiani, S., Sgoifo, A. and Mainardi, D. (1988b).Parental aggression displayed by female mice in relation to sex, reproductive status and infanticidal potential of conspecific intruders. Italian Journal of Zoology, 22, 193-201.

Perrigo, G. and Bronson, F. (1985). Sex differences in the energy allocation strategies of house mice. Behavioral Ecology and Sociobiology, 17, 297-302.

Perry, B., McCracken, A., Furr, B. and MacFie, H. (1979). Separate roles of androgen and oestrogen in the manipulation of growth and efficiency of food utilization in female rats. Journal of Endocrinology, 81, 35-48.

Petra, P. Stanczyk, F., Senear, D. Namking, P., Novy, M., Ross, J., Turner, E. and Brown, J. (1983). Current status of the molecular structure and function of the plasma sex steroid-binding protein (SBP). Journal of Steroidal Biochemistry, 19, 699-706.

Pointis, G., Latreille, M., Mignot, T., Janssens, Y. and Cedard, L. (1979). Regulation of testosterone synthesis in the fetal mouse testis. Journal of Steroidal Biochemistry, 11, 1609-1612.

Pointis, G., Latreille, M. and Cedard, L. (1980). Gonado-pituitary relationships in the fetal mouse at various times during sexual differentiation. Journal of Endocrinology, 86, 483-488.

Renoir, J., Mercier-Bodard, C. and Baulieu, E. (1980). Hormonal and immunological aspects of the phylogeny of sex steroid binding plasma protein. Proceedings of the National Academy of Science (USA), 77, 4578-4582.

Resko, J. (1975). Fetal hormones and their effect on the differentiation of the central nervous system in primates. Federation Proceedings, 34, 1650-1655.

Retzlaff, E. (1938). Studies in population physiology with the albino mouse. General Biology, 14, 238-265.

Reyes, F., Boroditsky, R., Winter, J. and Faiman, C. (1974). Studies on human sexual development. II. Fetal and maternal serum gonadotropin and sex steroid concentrations. Journal of Clinical Endocrinology and Metabolism, 38, 612-617.

Rines, J. and Vom Saal, F. (1985). Fetal effects on sexual behavior and aggression in young and old female mice treated with estrogen and progesterone. Hormones and Behavior, 18, 117-129.

Rohde-Parfet, K., Ganjam, V., Lamberson, W., Rieke, A., Vom Saal, F. and Day, B. (1988). Effects of intrauterine position on reproductive behavior and performance in gilts. Paper presented at the American Society of Animal Science meeting, New Brunswick, N.J.

Samuels, O., Jason, G., Mann, M. and Svare, B. (1981). Pup-killing behavior in mice : Suppression by early androgen exposure. Physiology and Behavior, 26, 473-477.

Selmanoff, M., Brodkin, L., Weiner, R. and Siiteri, P. (1977). Aromatization and 5alpha-reduction of androgens in discrete hypothalamic and limbic regions of male and female rats. Endocrinology, 101, 841-848.

Siiteri, P. and Thompson, E. (1975). Studies of human placental aromatase. Journal of Steroidal Biochemistry, 6, 317-322.

Siiteri, P. and Wilson, J. (1974). Testosterone formation and metabolism during male sexual differentiation in the human embryo. Journal of Clinical Endocrinology and Metabolism, 38, 113-125.

Siiteri, P., Murai, J., Hammond, G., Nisker, J., Raymoure, W. and Kuhn, R. (1982). The serum transport of steroid hormones. Recent Progress in Hormone Research, 38, 457-510.

Soares, M. and Talamantes, F. (1982). Gestational effects on placental and serum androgen, progesterone and prolactin-like activity in the mouse. Journal of Endocrinology, 95, 29-36.

Soares, M. and Talamantes, F. (1983). Midpregnancy elevation of serum androstenedione levels in the C 3H/HeN mouse : placental origin. Endocrinology, 113, 1408-1412.

Sodersten, P., Pettersson, A. and Eneroth, P. (1983). Pulse administration of estradiol-17β cancels sex difference in behavioral estrogen sensitivity. Endocrinology, 112, 1883-1885.

Southwick, C. (1958). Population characteristics of house mice living in english corn ricks : density relationships. Proceedings of the Zoological Society, London, 131, 163-175.

Svare, B., Kinsley, C., Mann, M. and Broida, J. (1984). Infanticide : Accounting for genetic variation in mice. Physiology and Behavior, 33, 137-152.

Sybulski, S. (1969). Testosterone metabolism by rat placenta. Steroids, 14, 427-447.

Toran-Allerand, C. (1984). On the genesis of sexual differentiation of the central nervous system : morphogenetic consequences of steroidal exposure and possible role of alpha-fetoprotein. Progress in Brain Research, 61, 63-98.

Vom Saal, F. (1976). Prenatal exposure to androgen influences intraspecific aggression in male and female mice. Doctoral thesis, Rutgers University.

Vom Saal, F. (1981). Variation in phenotype due to random intrauterine positioning of male and female fetuses in rodents. Journal of Reproduction and Fertility, 62, 633-650.

Vom Saal, F. (1983a). Models of early hormonal effects on intrasex aggression in mice. In : Hormones and Aggressive Behavior, Svare B. (Ed.), Plenum, New York, pp. 197-222.

Vom Saal, F. (1983b). Variation in infanticide and parental behavior in male mice due to prior intrauterine proximity to female fetuses : Elimination by prenatal stress. Physiology and Behavior, 30, 675-681.

Vom Saal, F. (1983c). The interaction of circulating estrogens and androgens in regulating mammalian sexual differentiation. In : Hormones and Behavior in Higher Vertebrates, Balthazart J., Prove E. and Giles R. (Eds.), Springer Verlag, Berlin, pp. 159-177.

Vom Saal, F. (1984). The intrauterine position phenomenon : Effects on physiology, aggressive behavior and population dynamics in house mice. In : Progress in Clinical and Biological Research, Vol. 169, Biological Perspectives on Aggression, Flannelly K., Blanchard R. and Blanchard D. (Eds.), Liss, New York, pp. 135-178.

Vom Saal, F. (1985). Time-contingent change in infanticide and parental behavior induced by ejaculation in male mice. Physiology and Behavior, 34, 7-15.

Vom Saal, F. (1988). Perinatal testosterone exposure has opposite effects on adult intermale aggression and infanticide in mice. In : House Mouse Aggression, Brain P., Mainardi D. and Parmigiani S. (Eds.), Horwood Academic Publishers, London, pp. 179-204.

Vom Saal, F. and Bronson, F. (1978). In utero proximity of female mouse fetuses to males : effect on reproductive performance during later life. Biology of Reproduction, 19, 842-853.

Vom Saal, F. and Bronson, F. (1980a). Sexual characteristics of adult female mice are correlated with their blood testosterone levels during prenatal development. Science, 208, 597-599.

Vom Saal, F. and Bronson, F. (1980b). Variation in length of the estrous cycle in mice due to former intrauterine proximity to male fetuses. Biology of Reproduction, 22, 777-780.

Vom Saal, F. and Finch, C. (1988). Reproductive senescence : Phenomena and mechanisms in mammals and selected vertebrates. In : Physiology of Reproduction, Knobil E. and Neill J. (Eds.), Raven Press, New York, pp. 2351-2413.

Vom Saal, F. and Howard, L. (1982). The regulation of infanticide and parental behavior : implications for reproductive success in male mice. Science, 215, 1270-1272.

Vom Saal, F. and Moyer, C. (1985). Prenatal effects on reproductive capacity during aging in female mice. Biology of Reproduction, 32, 1116-1126.

Vom Saal, F., Pryor, S. and Bronson, F. (1981). Effects of prior intrauterine position and housing on oestrous cycle length in adolescent mice. Journal of Reproduction and Fertility, 62, 33-37.

Vom Saal, F., Grant, W., McMullen, C. and Laves, K. (1983). High fetal estrogen titers correlate with enhanced adult sexual performance and decreased aggression in male mice. Science, 220, 1306-1309.

Vom Saal, F., Soares, M. and Johnson, D. (1987). Maternal stress reduces placental 17-hydroxylase and C17,2O lyase activity in mice (Mus musculus). Paper presented at the Society for the Study of Reproduction, Champaign.

Vom Saal, F., Even, M., Montano, M., Keisler, L. and Keisler, D. (1988). Maternal stress alters circulating testosterone and estradiol in male and female mouse fetuses : no correlation with postnatal traits. Paper presented at the Conference on Reproductive Behavior, Omaha.

Vomachka, A. and Lisk, R. (1986). Androgen and estradiol levels in plasma and amniotic fluid of late gestational male and female hamsters : Uterine position effects. Hormones and Behavior, 20, 181-193.

Vreeburg, J., Groenveld, J., Post, P. and Ooms, M. (1983). Concentrations of testosterone and androsterone in peripheral and umbilical venous plasma in fetal rats. Journal of Reproduction and Fertility, 68, 171-175.

Wade, G. and Gray, J. (1979). Gonadal effects on food intake and adiposity : a metabolic hypothesis. Physiology and Behavior, 22, 583-593.

Weisz, J. and Ward, I. (1980). Plasma testosterone and progesterone titers of pregnant rats, their male and female fetuses, and neonatal offspring. Endocrinology, 106, 306-316.

Welshons, W. and Gorski, J. (1986). Nuclear location of estrogen receptors. In : The Receptors, Vol. IV, Conn P. (Ed.), Academic Press, New York, pp. 97-147.

Westphal, U. (1986). Steroid-protein interactions. II. Springer Verlag, Berlin

Wilson, J., George, F. and Griffin, J. (1981). The hormonal control of sexual development. Science, 211, 1278-1284.

Yasukawa, N., Monder, H., Leff, F. and Christian, J. (1985). Role of female behavior in controlling population growth in mice. Aggressive Behavior, 11, 49-64.

4 Heterotypical Sexual Behaviour in Female Mammals

Robert W. Goy and Marc Roy *

The patterns of sexual behaviour that are appropriate and necessary for reproduction are relatively fixed within a species and distinct for each sex. In general, males display responses that serve the function of attracting or locating a female partner, responses that function to court or persuade the partner, and responses that facilitate sexual union such as mounting, intromission, and ejaculation. Females, at all stages of this enterprise, display reciprocity of responses such that reduction of the distance between herself and the male partner is facilitated or permitted, encouragement or stimulation of the male's sexual advances is provided (proceptive behaviour), and accomodation is made to sexual union (receptive behaviour). The diversity of responses displayed to accomplish these ends is very great among species, and the degree to which males display responses that differ from those of females also varies greatly. Nevertheless, within a species the responses subserving each function occur regularly enough and are closely enough associated with one of the sexes that it is possible to identify both male-typical and female-typical behavioural repertoires that ensure reproductive success. When males and females are displaying responses usually associated with the reproductive activities of their own sex, the display is characterized as "homotypical". When a male or female displays behaviour normally associated with the repertoire of the opposite sex, the display is characterized as "heterotypical". Mounting is by far the most commonly displayed heterotypical behaviour in female mammals, and the current report focuses primarily on this particular activity.

Since the work of Josephine Ball and William Young and his

* Wisconsin Regional Primate Research Center, University of Wisconsin, 1223 Capital Court, Madison Wisconsin 53715-1299, USA.

colleagues beginning in the 1930' s, it has been apparent that steroid hormones have the ability to augment mounting activity in adult females. Considerable experimentation has been undertaken to elucidate which hormones can regulate mounting behaviour to determine how hormones during early development, during adulthood, or during both developmental stages contribute to the expression of mounting by females. A singularly instructive review was provided by Beach (1968) in which not only hormonal influences, but also an assessment of environmental requirements was provided. In this chapter we will consider these areas of research in several groups of mammals that have been extensively studied : rodents, carnivores, and primates.

Display of Heterotypical Mounting Behaviour by Normal Females

Since mounting behaviour is seen in many female mammals during oestrus, it is natural to ask whether this behaviour is regulated by ovarian hormones. If mounting is abolished by ovariectomy, or if it shows variability in frequency of occurrence across the ovarian cycle, then hormones such as oestrogen and progesterone may be important in the activation of this behaviour. Mounting behaviour displayed by males of many species is to varying degrees modulated by testosterone, and, one can ask if this androgen can activate mounting in females. The treatment of normal adult females with androgen allows a more complete assessment of the bisexual potential of them than would be possible by any other available means.

In the guinea pig, mounting typically is seen only during proestrus and early oestrus (Young & al., 1935), and females will pursue and mount both males and females (Young & Rundlett, 1939). Mount frequency is usually low during the first heat that a female shows, and, the frequency (rate) increases with age (Young & al., 1939). Inasmuch as the display of mounting was found to be associated strongly with the display of receptive behaviour, it was postulated that the synergistic actions of oestrogen and progesterone were responsible for both receptivity and mounting (Young & al. 1939). Subsequently, Young and Rundlett (1939) demonstrated that spayed females given sequential injections of oestrogen and progesterone separated by 36-48 hours were the most effective in eliciting both mounting and lordosis (the postural response of the female that indicates receptivity). Injections of oestrogen alone induced mounting in less than 15 % of the females and injections of progesterone without prior oestrogen conditioning failed to elicit either receptivity or mounting. Similarly, when spayed and tested for mounting behaviour without hormone replacement females displayed very low mount frequencies (Goy & al., 1964) or did not mount at all (Phoenix & al., 1959). The inference, based on this work, that mounting in the normal cycling guinea pig, is regulated by exposure of the brain to ovarian oestrogen and to progesterone some hours later is supported by direct evidence of a a preovulatory surge of progesterone in this species (Feder & al., 1968).

Gonadectomized female guinea pigs also display mounting behaviour when treated with daily injections of testosterone (Phoenix & al., 1959 ; Gerall, 1966 ; Goldfoot, 1979). Although mounting frequencies are increased with this treatment, they are less than those displayed by males treated with testosterone or females treated with oestrogen and progesterone (Phoenix & al., 1959). Interestingly, neither oestrogen (Young & Rundlett, 1939), dihydrotestosterone (Goldfoot, 1979) nor the combination of oestrogen and dihydrotestosterone (Goldfoot, 1979) will stimulate mounting behaviour in spayed females. These findings could lead to the conclusion that neither of the major metabolites of testosterone are necessary for the androgenic activation of mounting behaviour. However, recent experiments using an aromatase inhibitor to prevent aromatization of testosterone to oestrogen revealed that oestrogens probably are necessary for the activation of mounting behaviour by testosterone (Roy & Goy, 1988).

These descriptions of the actions of steroids in the guinea pig are based on studies conducted in an outbred (Topeka) stock. Females from one known inbred strain do not respond in the same manner to these hormones. Strain 2 females rarely mount either during oestrus or when treated with exogenous testosterone (Wallen, 1978). Strain 13 females in heat mount at much higher frequencies than either Topeka or Strain 2 females (Goy & Young, 1957). Thus, in the guinea pig, genetic factors also contribute to potential of females to display mounting behaviour (Goy & Jakway, 1959).

In the female rat, unlike the guinea pig, early studies by Beach indicated that mounting is independent of ovarian secretions. Beach (1942) demonstrated that intact females mount and that ovariectomy did not reduce the frequency of mounting. However, the frequency did vary across the cycle, with a slight reduction during oestrus (Beach & Rasquin, 1942). Sodersten (1972) also found that mounting frequencies decreased during oestrus. Furthermore, he noted that they were decreased but not abolished following ovariectomy. Oestrogen replacement restored the mount frequencies to intact levels. Similar findings have been reported by Pfaff (1970) and De Jonge et al. (1986). During oestrus, female rats are most receptive to mounts from female partners, and Södersten (1972) has suggested that the mounting by partner females reduces the probability that the oestrus female herself will mount. Alternatively, the decrease in mounting during oestrus may be due to partial inhibition by ovarian progesterone (Sodersten, 1972) which is secreted maximally during oestrus (Feder & al., 1968). Thus, inasmuch as mounting behaviour persists following ovariectomy, but varies across the cycle, it appears that ovarian hormones (oestrogen and progesterone) act more to modulate the frequency of mounting than to initiate mounting.

Androgens are also capable of increasing mount frequencies in intact (Ball, 1937, 1940) and ovariectomized (Beach, 1942 ;

Södersten, 1972) rats. Whether testosterone or one of it's metabolites is stimulating mounting is not clear. As mentioned above, oestrogens can increase mount frequencies in spayed rats. Gladue (1984) has demonstrated that another metabolite, 5α -dihydrotestosterone, also increases mount frequencies in ovariectomized females. Thus, any, or all of these hormones may be working singly or together to augment the display of mounting above that characteristic of spayed rats.

Until recently experimental attempts to stimulate mounting behaviour in female hamsters have been unsuccessful. Early reports indicated that mounting was not displayed at any time during the oestrus cycle. Moreover, neither ovariectomized nor testosterone-treated females displayed mounts (Tiefer, 1970 ; Carter & al., 1972 ; Paup & al., 1972). However, Hsu and Carter (1986) reported contradictory findings. Female hamsters that were group-housed showed robust mounting responses when they were in oestrus or ovariectomized and treated sequentially with oestrogen and progesterone. Methodological differences may account for the difference in the reported results. In the study done by Tiefer (1970), the subjects were housed individually and paired briefly with a receptive female partner. According to Hsu and Carter (1986) singly caged females rarely display mounting. The studies done by Paup & al. (1972) and Carter & al. (1972) were studies of the effects of prenatally administered androgens on the display of adult sexual behaviour. In these studies normal females (those receiving no prenatal androgens) did not mount as adults either prior to or following testosterone administration in adulthood. However, in the tests given prior to testosterone treatment (Carter & al., 1972), no mention was made of whether the females were in oestrus. All of the pre-androgen tests done by Paup & al. (1972) were given following ovariectomy and without oestrogen or progesterone treatment. Thus, it appears as if in these studies the appropriate conditions (hormonal and/or housing) were not adequate to permit mounting displays. It is interesting that unlike the rat or guinea pig, female hamsters do not display mounting after testosterone administration, even when group housed. This species difference cannot be easily accounted for, and explanations must await further research.

Among carnivores, the two species in which hormonal actions on female mounting have been studied best are the ferret and dog. Male ferrets typically only mount once prior to ejaculation. Therefore, mounting is quantified by measuring the duration of a mount rather than the frequency of mounting (Baum, 1976). Oestrus female ferrets do mount, although the duration is considerably less than in males (Baum, 1976). Following ovariectomy, mounting is abolished and is not restored by treatment with oestradiol or testosterone (Baum, 1976 ; Baum & al., 1982a). From these studies it is not readily apparent which hormones are responsible for the mounting seen during oestrus.

In contrast to the ferret, female dogs display mounts regardless of their own hormonal status, and when either in an oestrus or anoestrus condition they will mount other females. Adult bitches, even if ovariectomized prepubertally, will also mount female partners. However, the observed frequency of mounting is greatest when the partner female (rather than the actor female) is in oestrus (Beach & al., 1968). Testosterone treatment has no effect on the mount frequencies displayed (Beach & al., 1972).

Female dogs will also mount male partners when the female is in oestrus. In limited, non-systematic tests with anoestrus females no mounting of male partners was seen (Beach & al., 1968). These observations are interesting and suggest that mounting of female partners is not normally regulated by ovarian steroids, but the mounting of male partners is so regulated. However, more precise studies are needed to assess whether female mounting of males is hormonally regulated.

There is also variation within primate species in the degree to which mounting by females is influenced by hormones. In rhesus macaques (Macaca mulatta) mounting is shown even by juvenile (prepuberal) females, although the proportion of mounting females is not great. The factor that appears to be most related to mounting frequency is the dominance status of an individual. Females of high rank mount much more frequently than do females of lower rank. These patterns do not change with the onset of puberty suggesting that mounting of peers by juvenile females is not related to either the mounter's of mountee's endocrine state.

In the adult female rhesus there is some suggestion that female mounting is related to the ovarian cycle. In a study of two females, Michael & al. (1974) noted that female mounting of male partners occurred most frequently close to the time of expected ovulation. Following ovariectomy there was a significant decline in mounting behaviour which was restored by oestrogen therapy. Female-female mounting also may be influenced by oestrogen. Akers and Conaway (1979) reported that the mounting female was usually in the follicular or periovulatory phase of the cycle. Furthermore, Pope & al., (1987) noted that ovariectomized females treated with oestrogen mounted other females. While these studies suggest that ovarian oestrogen may regulate mounting of both male and female partners, this conclusion must be viewed cautiously. Only small numbers of females have been studied (Michael & al., 1974 ; Akers & Conaway, 1979) and Pope & al. (1987) do not mention whether mounting occurred prior to or following oestrogen administration.

In the rhesus macaque, mounting by females is not stimulated by testosterone. Administration of as much as 2 mg/kg/day of TP to ovariectomized females for 12 weeks did not result in the display of mounting behaviour (Pomerantz & al., 1986).

In the stumptail macaque (Macaca arctoides) mounting by females appears to be independent of high oestrogen levels. In 23 recorded occurrences of female-female mounting, the mounter was never in, or within 2 days of oestrus. In all cases the mounting female was lactating (Chevalier-Skolnikoff, 1976). In many cases,

the female who was mounted had previously engaged in heterosexual encounters. Therefore, it is not possible to determine in this
case whether mounting is hormonally or socially mediated or if
both factors contribute to the expression of female mounting.

As in stumptail macaques, mounting in Japanese macaques
(Macaca fuscata) appears to be independent of cyclic ovarian secretions. However, the length of time since parturition and the
occurrence of pregnancy may influence the display of female-male
mounting. Females who had not given birth during the previous
year were more likely to mount male partners than females who
had recently given birth (Fedigan & Gouzoules, 1978). Female-male
mounting also appears to occur more frequently following
conception (Gouzoules & Goy, 1983). On the other hand, female-
female mounting was not associated with conception. That is,
females who mounted other females were no more likely to subsequently give birth than those females who did not engage in
mounting (Gouzoules & Goy, 1983). It should be pointed out that in
these studies no hormonal measurements were obtained. Thus while
conditions of pregnancy or the length of time since parturition are
related to mounting, which, if any, hormones are involved is not
known.

Among the great apes, mounting behaviour by adult females is
not a regular occurrence. Indeed, Dagg (1984) has characterized it
as "occasional" or "rare" both in captivity and the wild based on
observations of chimpanzees by Yerkes (1939) and Van Lawick-
Goodall (1968). Furthermore, little is known about possible
hormonal influences on female heterotypical behaviour in this great
ape species, and no experiments have been done to evaluate the
possibilities. Recent observations of a large captive group of
bonobos (the so-called pygmy chimpanzee, De Waal, 1989) portray a
totally different picture. In this species of great ape female
homosexual behaviour is displayed frequently and by females of all
ages. Accordingly, specific endocrine conditions are not required
for such activities, and it is doubtful that the behaviour is even
influenced by hormones. It should be pointed out, however, that
this species makes extensive use of the sexual behaviour repertoire
for nonreproductive functions, and such uses are not limited to
females. Males of all ages behave similarly, and De Waal has
estimated that "plus de trois-quarts des comportements sexuels et
érotiques observés n'avaient aucune fonction reproductrice..." (De
Waal, 1989). Sexual activities, both homotypical and heterotypical,
occur with regularity in the context of social tension and seem to
function to reduce such tension.

When female mounting is associated with oestrus, its frequency
of display by spayed females can be augmented by a single
injection of oestrogen (or progesterone). It should be noted that in
many species the amounts of hormone that must be administered to
augment mounting are much greater than the amounts normally
produced by the ovaries or the ovaries and the adrenals together.

This is particularly true of those treatments involving either the oestrogens alone or the androgens. When female mounting is associated with oestrus, its frequency of display by spayed females can be augmented by a single injection of oestrogen (or by sequential single injections of oestrogen and progesterone). Effective treatments with these hormones often parallel or duplicate the requirements for augmentation of mounting in the castrated male. Viewed in this manner, testosterone-activated mounting in females is a masculine homologue, and the effectiveness of testosterone in augmenting such behaviour may be regarded as an index of the bisexual potential of the individual or species.

It is rare that the bisexual potential of females is so great that their hormonal requirements for the expression of mounting are either equal to or less than those of males. Indeed, the common finding is that females generally require more testosterone than males to display similar frequencies of mounting. However, this inequality of the sexes can be reduced or eliminated by exposing females to androgens early in development in addition to treatment with testosterone in adulthood. In 1959, Phoenix et al. demonstrated that testosterone can act prenatally to permanently alter the expression of mounting such that it is exhibited in a "male-typical" pattern. The effects described by Phoenix et al. (1959) have been termed "organizational" to distinguish them from the transitory and reversible "activational" effects of hormones usually associated with hormonal influences in adulthood.

Augmentation of Heterotypical Behaviour by Hormonal Intervention Early in Development

As we have discussed, the heterotypical behaviours displayed by normal females are mostly limited to mounting. However, male typical sexual behaviour is more extensive, consisting of precopulatory behaviours (courtship and inter-male aggression) as well as copulatory behaviours (mounting, intromission, and ejaculation). When females are exposed to testosterone or, in some cases oestradiol, during early stages of development, they are more likely to display mounts and other heterotypical sexual behaviours than are untreated females. Moreover, the hormonal requirements for the display of these behaviours often more closely parallels that of males. The period of time during which steroids can have these influences varies among species from during gestation (prenatal) to just after birth (neonatal).

Phoenix & al. (1959) were the first investigators to demonstrate that female heterotypical behaviour can be augmented by androgens given prior to birth. They noted that female guinea pigs treated prenatally with testosterone displayed mounting frequencies as adults that were greater than those of control females both prior to and after the administration of testosterone in adulthood. Furthermore, the mounting frequencies displayed by prenatally androgenized females during adult testosterone treatment were comparable to those displayed by males. As pointed out earlier, females treated only in adulthood with testosterone display augmented mounting frequencies, but mount less frequently than males. Prenatal testosterone administration also results in increased frequencies of intromissions (Goldfoot & Van der Werff ten Bosch, 1975) and occasionally in the display of ejaculatory behaviour (Gerall, 1966). Increases in heterotypical mounting behaviour of female guinea pigs are also seen with the prenatal injection of the synthetic oestrogen diethylstilboestrol (DES) (Hines & Goy, 1985 ; Hines & al., 1987) and the neonatal injection of oestradiol dipropionate (Feder & Goy, 1983). In the latter treatment, however, the augmentation of mounting is restricted to the display associated with oestrus.

The suggestion that prenatal oestrogens contribute to the development of mounting behaviour in female guinea pigs is further strengthened by evidence from Hines & al. (1987). They noted that adult females treated with the antioestrogen tamoxifen in utero showed diminished mounting behaviour in the absence of hormonal stimulation as well as following treatment with oestradiol and pro-gesterone or with testosterone. Thus, in the guinea pig, both androgens and oestrogens may be necessary for the development of the potential to show mounting behaviour.

There is some disagreement as to whether perinatal androgen injections (those immediately prior to and/or immediately after parturition) lead to augmented mounting frequencies in adult female rats. Södersten (1973) and Christensen and Gorski (1978) found that neonatal androgen administration did lead to an increase in the display of mounting behaviour, while Pfaff and Zigmond (1971) and Whalen & al. (1969) were unable to demonstrate this. Evidence supporting the augmentation of mounting in female rats by prenatal hormonal influences comes from another line of work by Clemens and his collaborators. Females that were located next to males in utero were found to display more mounting behaviour when administered testosterone in adulthood than did females that were not in close proximity to male foetuses. It has been suggested that the increased potential to mount is due to androgens from the male foetuses masculinizing the females (Clemens & Coniglio, 1971 ; Clemens & al., 1978).

There is little doubt that perinatal androgenization leads to an increase in copulatory behaviours other than mounting. Increases in intromission frequency were noted by Whalen et al. (1969) and both intromissive and ejaculatory behaviours in females have been documented by Ward (1969), Sachs et al. (1973), and Sachs and Thomas (1985). It could be argued that the increase in copulatory behaviours seen in perinatally androgenized females is due to the presence of a male like phallus, and not to a change in those areas of the brain that mediate masculine reproductive behaviours. However, Christensen and Gorski (1978) administered either testosterone or oestradiol directly into the hypothalamic nuclei of neonatal rats and found increased mounting and intromissive-like behaviour. In this case actual intromission was not possible due to the lack of a phallus, but the deep thrusts characteristic of intromission were seen. Thus, in the female rats, androgens can act during certain critical periods of development to increase the display of heterotypical copulatory behaviour.

Mounting has only been observed in oestrus female hamsters living in groups, and testosterone has not been shown to activate mounting behaviour in normal adults. This deficiency can be overcome by neonatal testosterone (Swanson, 1971 ; Paup & al., 1972 ; Carter & al., 1972) or diethylstilboestrol injections. Paup & al. (1972) demonstrated that females given testosterone or diethylstilboestrol injections shortly after birth and treated with testosterone as adults displayed mounting behaviour. When adult testosterone administration ceased, the mounting frequencies gradually returned to zero. This suggests that androgens and oestrogens act in the neonatal period to alter the sensitivity to testosterone in adult females, and that early hormone exposure is not sufficient to induce hormonally independent mounting. However, there are some data that may conflict with the second conclusion. Carter & al. (1972) also found that neonatal injection of testosterone augmented mounting behaviour. Contrary to the findings of Paup & al., this was observed in females prior to and following ovariectomy as well as following adult testosterone treatment. That is, the frequency of mounting was not altered by the adult endocrine status. The discrepancies between these two studies may be due to the high levels of mounting seen by Carter & al. (1972) while the ovaries were present. Mounting behaviour decreases slowly after castration in males and it is possible that insufficient time was allowed following ovariectomy to see such a decline in these females. Subsequent testosterone administration had little influence, apparently because the females were mounting at or near maximal rates. Thus, the finding of Paup & al. (1972) that mounting in neonatally androgenized females is dependent on adult androgen is not necessarily at odds with the results of Carter & al. (1972).

Mounting is rarely seen in adult female ferrets, regardless of the endocrine status. Similarly, neck-gripping, another male-typical sexual behaviour is rarely seen. As in several other species, neonatally administered testosterone greatly augments the display of these behaviours (Baum, 1976 ; Baum & al., 1982a). However, the display of neck-gripping and mounting is contingent on the presence of testosterone or oestradiol in the adult animal. In contrast, neonatal oestrogen treatment resulted in only marginal increases in heterotypical behaviour.

The female dog will normally mount another female, although intromissive-like behaviour is not seen. Moreover, treatment of normal adult bitches with testosterone does not augment this behaviour. Both the percentage of tests that are positive for mounting and the mount frequency are augmented when females are treated prior to and following birth with testosterone. Furthermore, females that have received androgens both in utero and during the neonatal period will display intromissive-like thrusts during adult testosterone treatment, although actual insertion of the phallus has not been observed (Beach & al., 1972).

Much work over the past 15 years has served to demonstrate that androgens influence the development of behaviour during the prenatal period in the Rhesus monkey (Goy, 1978, 1981). When genetic female Rhesus monkeys are exposed transplacentally to exogenous testosterone during the latter two-thirds of gestation, their early social and protosexual behaviour can be completely altered so that predominantly heterotypical behaviour patterns are displayed. The display of these behaviour patterns (augmented mounting of peers, play initiation, and rough play) is not dependent upon any postnatal action of hormones and the effects of the prenatal treatment are permanent (Goy & Kemnitz, 1983). The magnitude of the psychosexual change as well as its specific nature are completely and uniquely determined by the timing, amount and duration of exposure to testosterone prenatally. Moreover, it has been shown recently that the behaviour of females can be masculinized by timing their exposure to testosterone in such a way that no virilization of the genital structures is induced (Goy & al., in press). Still other specific, but different treatments are associated with virilization of the genitalia and a lack of any behavioural masculinization.

As adults receiving concurrent testosterone treatment, females treated prenatally with androgens, unlike normal females, showed increased mounting frequencies although they mounted less than intact males. Furthermore, they displayed purse-lip gestures (a male typical courtship display) at frequencies indistinguishable from males. These increases have been noted when the prenatally andro-genized females were tested with either one (pair test) or two (trio test) stimulus female partners (Pomerantz & al., 1986 ; Pomerantz & al., 1988). When rhesus males are observed in trio tests, they display a strong preference for one of the two partner females and direct most of their courtship and sexual behaviours to that female. Prenatally androgenized females, but not normal females, also display this kind of partner preference. Thus, in the rhesus monkey treatment of females with androgens prior to birth increases the amount of male-typical courtship as well as the amount of male-typical copulatory behaviour that will be shown when the animals reach adulthood.

We have provided evidence that in several species both androgens and oestrogens can act early in development to augment adult heterotypical behaviour in females. In some species the biological conversion of testosterone to oestradiol (aromatization) within developing neural target cells may be required for this process. All of the evidence for this requirement, however, is specific to non-primate mammals. There are good reasons to hypo-thesize that oestrogens do not have these behavioural effects in primates. For example, the psychological characteristics of the tfm rat and the androgen-insensitive human being are totally different even though the genetic disorder produces homologous disturbances in androgen action in both species. The tfm male rat, phenotypi-

cally female, shows masculine sexual behaviour when castrated and
treated with androgens or oestrogens in adulthood (Olsen, 1979).
These males are also defeminized and show frequencies of female-
typical receptive postures that are similar to wild-type males, and
significantly lower than those of female littermates (Shapiro & al.,
1980 ; Olsen & Whalen, 1981). This male-typical sexual orientation
has been attributed to actions of oestrogen derived from the in situ
aromatization of androgen produced by their testes. In contrast,
the androgen-insensitive (AI) human male, also phenotypically
female, has a completely feminine psychosexual orientation and
life-style (Money & al., 1967). In the AI human male, the testes
produce   normal   or   supernormal   amounts   of   testosterone.
Aromatization is not deficient, and as adolescents these afflicted
individuals produce enough oestrogen derived from testosterone to
cause development of female breasts. The responsiveness of the
tissues comprising female secondary sexual characteristics is an
important demonstration that AI males are not insensitive to
oestrogens, only to androgens. Assuming that the machinery for
aromatization existed during foetal life, the derived oestrogens
could be expected to cause psychological masculinization according
to predictions based on lower mammals. Clearly, however, this does
not happen in the human being. Yet psychosexual virilization by
excessive prenatal androgen does occur in human beings in the
Congenital Adrenal Virilizing Syndrome (Money & Schwartz, 1977).
The inference to be drawn from this set of findings is that
primates (as represented by human clinical findings) cannot be
psychosexually virilized by oestrogens derived from androgens, but
require the actions of androgens directly on the cellular machinery.

Another line of inquiry into the question of actions of prenatal
oestrogens on human females has been carried out by studying the
female offspring of mothers treated during pregnancy with DES. A
large number of young women were exposed to this drug during the
1940s, 1950s, and 1960s. There is some evidence to suggest that
DES can cause augmentation of masculine characteristics. Hines
(1982) reported this in a study of cognitive and spatial abilities of
exposed women. Ehrhardt and her collaborators (1985) have
reported an increased incidence of homosexuality (masculinized
sexual orientation) in women exposed to DES prior to birth. However, neither of these studies are prospective and developmental
social factors that may have contributed to these alterations
cannot be assessed. Thus the results do not permit the definitive
conclusion that DES is the real cause of the masculinized sexual
orientation.

Only one study has been reported on prenatal oestrogen and
postnatal behaviour of non-human primates. This study was carried
out by Golub et al. (1983) who administered Norlestrin (Parke -
Davis) to pregnant rhesus mothers during the first third of pregnancy. No disorders of social behaviour and no augmentations of
heterotypical behaviour were found. It should be noted, however,
that the treatments evaluated were given during a developmental
period when even testosterone has no masculinizing effects on
behaviour (Goy, 1981).

What seems likely from this brief survey is that oestrogens before birth do not cause psychosexual virilization in primates. This inference is strengthened by the finding that dihydrotestosterone (a non-aromatizable androgen) is nearly as effective as testosterone (which is aromatizable) in increasing the heterotypical behaviour of rhesus females (Goy, 1981 ; Pomerantz & al., 1985). However, until prospective, controlled studies are carried out in primates, the role that oestrogens might play in the patterning of heterotypical behaviour will be unknown.

Neural Regulation of Mounting

While it is widely accepted that gonadal steroids act on the brain to influence the expression of sexual behaviour, there is a paucity of information on the neural regulation of heterotypical behaviour in female mammals. There has been, however, intensive investigation of the neural basis of homotypical copulatory behaviour in males of several species. It is possible that the information gained from studies in males may be useful in understanding some of the mounting responses of females. Investigations of which brain structures are involved in the mediation of mounting behaviour have taken several forms including lesions, anatomical localization of sexually dimorphic, steroid binding regions, and direct application of hormones to particular brain areas. Much research directed at the hypothalamus and preoptic area has shown that these areas are critical for the expression of mounting behaviour. However, other brain sites, both subcortical and cortical, are also involved in the mediation of mounting.

Many areas of the mammalian brain, including the medial preoptic area, hypothalamus, amygdala, and the stria terminalis can be labelled with tritiated steroids (Pfaff, 1968 ; Stumpf, 1970 ; Krieger & al., 1976 ; Warembourg, 1977 ; Rees & Michael, 1982). Furthermore, androgen and oestrogen receptors have been localized in these and other areas in adult males and females of several species (McEwen, 1980 ; Rainbow & al., 1982 ; Bonneau & al., 1987). The brains of foetal and neonatal males and females also contain receptors for both of these hormone classes (MacLusky & al., 1979a, b ; Vito & Fox, 1982 ; Pomerantz & al., 1985). Thus it is not surprising that these areas have been implicated in the mediation of mounting.

Lesions of the medial preoptic area-anterior hypothalamus (MPOA-AH) in males lead to a near complete loss of copulatory behaviour in the guinea pig and rat (Brookhart & Dey, 1941 ; Phoenix, 1961 ; Heimer & Larsson, 1966, 1967 ; Lisk, 1968 ; Giantonio & al., 1970 ; Ginton & Merari, 1977). These lesions were most effective when large and encompassing both the MPOA and AH. Similar findings have been demonstrated in the male dog (Hart & Ladewig, 1979) and rhesus macaque (Slimp & al., 1978). In adult female rats that were receiving testosterone, MPOA and AH lesions reduced the number of mounts displayed (Singer, 1968 ;

Gray & Brooks, 1984). These studies would indicate that the MPOA-AH is an important locus for the display of male-typical sexual behaviour in both males and females. However, within the hypothalamus, there may be more than one area involved in the mediation of hormonally activated mounting. Goy and Phoenix (1963) noted that ovariectomized female guinea pigs with large hypothalamic lesions failed to display normal receptivity and mounting when given sequential injections of oestradiol and progesterone. These same females, however, displayed mounting when treated with testosterone.

Direct application of gonadal steroids to the hypothalamus and preoptic area has also suggested that these areas are important for the hormonal activation of mounting. In the castrated male rat, copulatory behaviour can be activated by placing testosterone containing cannulae in the preoptic area (Davidson, 1966 ; Christensen & Clemens, 1974) or the anterior hypothalamus (Davidson, 1966 ; Johnston & Davidson, 1972). All components of copulatory behaviour can also be activated when oestrogen is applied to the hypothalamus (Christensen & Clemens, 1974). In the guinea pig, when dihydrotestosterone (which activates mounting when given systemically to males) was applied to the hypothalamus, it activated all components of copulatory behaviour (Butera & Czaja, 1985). In the castrated male hamster, as in the male rat, either testosterone or oestrogen containing cannulae directed at the hypothalamus were found to activate mounting (Lisk & Bezier, 1980), but in order to elicit intromissions and ejaculation, these treatments had to be accompanied by systemically administered androgen (Lisk & Greenwald, 1983). Thus, the hormones that are known to elicit sexual behaviour when administered systemically in several male rodents also activate many of the same behaviours when their site of action is confined to the hypothalamus.

Morphometric sexual dimorphisms have been described in the MPOA-AH which also suggest a role of this area in the regulation of sexual behaviour. Raisman and Field (1973) described a sex difference in the synaptic organization in the rat preoptic area. Sex differences in patterns of dendritic branching have been reported in both the hamster (Greenough & al., 1977) and macaques (Ayoub & al., 1983). In many species differences in the cytoarchitecture of nuclei within the MPOA-AH have also been seen. Perhaps the most notable difference is in the volume of the sexually dimorphic nucleus of the preoptic area (SDN-POA). This nucleus is considerably larger in male rats than in females (Gorski & al., 1978 ; Gorski & al., 1980). Similar nuclei have been described in the Guinea pig, mouse, hamster, gerbil and ferret (Bleier & al. 1982 ; Commins & Yahr, 1984 ; Hines & al., 1985 ; Tobet & al., 1986 ; Byne & Bleier, 1987). In nearly all cases hormonal treatments early in development that change the behaviour of females to a male-typical pattern also increase the volume of this nucleus. Thus, treatment of female rats perinatally with TP (Gorski & al., 1978 ; Döhler & al., 1982 ; Döhler & al.,

1984) or DES (Döhler & al., 1984) increased the size of the
SDN-POA such that it was comparable to that of males. In the
female guinea pig administration of DES prenatally to females
(Hines & al., 1987) increased both the display of mounting
behaviour and the volume of the SDN-POA. Byne and Bleier (1987)
describe similar results with TP administration. However, in one
treatment group they were able to increase mounting behaviour of
females without altering neural structures. The authors suggest
that the sex differences in the volume of the medial preoptic
nucleus are not sufficient to account for the sex differences in
reproductive behaviour. A similar conclusion can be reached from
work with brain lesions. While large lesions in the MPOA-AH
disrupt male-typical copulatory behaviour, discrete lesions of the
SDN-POA in male rats do not result in deficits in sexual
performance (Arendash & Gorski, 1983). Thus, it appears that
although changes in the organization of the MPOA are correlated
with changes in sexual behaviour, other nuclei must be involved in
the regulation of the expression of copulatory behaviour.

Since Klüver and Bucy's (1939) observation of hypersexuality in
rhesus monkeys following ablation of the temporal lobe and underlying limbic structures, many studies have examined the effects of
such lesions. These have focused primarily on the amygdala and
stria terminalis which send projections to the preoptic area and
hypothalamus (De Olmos, 1972 ; Swanson, 1976). Kling (1968)
reported that large lesions of the amygdala in juvenile male rhesus
increased the frequency of normally oriented mounting. Furthermore, mounting was often solicited by males and accompanied
by an erection in the male being mounted. This apparent
hypersexuality, however, may be limited to the male rhesus in the
laboratory setting. In a semi-free ranging environment, males with
amygdaloid lesions withdrew from their social groups and did not
exhibit any sexual behaviour (Dicks & al., 1969). Furthermore,
laboratory-housed adult female rhesus macaques with lesions of the
amygdala did not show any changes in their sexual behaviour when
tested with adult male partners (Spies & al., 1976).

In male rodents, amygdaloid lesions have a different effect on
sexual behaviour than in the male rhesus. In both the rat (Giantonio
& al., 1970) and hamster (Lehman & al., 1980) lesions of the corticomedial nucleus of the amygdala result in impaired copulatory
performance. In the rat, this was characterized by an increased latency to ejaculation, whereas in the hamster, copulatory behaviour
was rarely seen following lesioning. These deficits in copulatory behaviour are site specific because, in both species, lesions of the
basolateral amygdaloid nucleus did not result in any behavioural
changes.

Using a different technique, Baum & al. (1982b) also have
implicated the medial amygdala in the neural mediation of
copulatory behaviour in the male rat. They observed that DHTP
pellets implanted into the medial amygdala of castrated rats that
also received small amounts of oestrogen systemically increased
both intromissions and ejaculations beyond the level seen in males

without DHTP implants.

The corticomedial nucleus of the amygdala projects via the stria terminalis to the bed nucleus of the stria terminalis which then sends projections to the medial preoptic area (De Olmos, 1972 ; Swanson, 1976). In the male rat, lesions of the stria terminalis, like lesions of the corticomedial amygdala and the medial preoptic area, result in deficits in copulatory performance. Giantonio et al. (1970) found that electrolytic lesions of the stria terminalis led to an increase in the latency to ejaculation. Using a knife cut procedure, Paxinos (1976) noted a more severe reduction of copulatory activity : the frequency of mounts, intromissions and ejaculations all decreased.

It appears then that disruption of the pathway from the amygdala to the MPOA results in decreased copulatory behaviour in the male. However, the specific effects vary with the site of the lesion, the species, and the sex studied, as well as the environment in which the animals are observed.

Lesion studies have demonstrated that the cerebral cortex is also an important area for the display of mounting behaviour. Electrocauterization of large amounts of the cortex abolished mounting in both the male rat (Beach, 1940) and female rat (Beach, 1943). The decrease in mounting behaviour was proportional to the amount of cortex removed in both sexes and lesions of less than 20 % of the cortical surface did not affect the behaviour. Furthermore, the decrease in behaviour was independent of the location of the lesion. In a more detailed examination of the behavioural changes, Larsson (1962, 1964) found that lesions encompassing less than 37 % of the frontal lobes completely abolished mounting in nearly half of the male rats tested. Similar lesions of the posterior lobe resulted in minor reductions in behaviour and were attributed to visual deficits (Larsson, 1964).

From this brief survey, it appears that a number of neural sites are important for the display of mounting and other copulatory behaviours in male mammals. It is still unclear whether all of these areas are also necessary for the display of mounting in the female. While lesions of the MPOA and cortex disrupt mounting behaviour in both sexes, there is not sufficient information to determine whether the same is true of lesions in the amygdala and stria terminalis. Further studies also need to examine how these structures interact to regulate mounting and whether all of these structures are involved in all types of mounting (i.e. non-hormonally mediated, E+P mediated, and androgen mediated).

It is also important to note that in the species that we have discussed, androgen and oestrogen uptake in the brain has been demonstrated autoradiographically and receptors for these steroids have been found. Even if the receptors are located in the MPOA, and this area is implicated by lesion studies as being important for mounting, this does not guarantee that a female will mount when treated with a particular steroid. Either these approaches do not offer a fine enough analysis to tell which particular sites are important for the regulation of mounting or merely having steroid uptake and receptors in particular areas is not sufficient for the display of mounting. How the steroid/receptor complexes interact with the genome in particular neurons is probably of critical importance.

DISCUSSION

Mounting behaviour exhibited by normal females can be placed into three broad classes which are distinct in their endocrine regulation : (1) mounting that is free of ovarian influences and persists following gonadectomy ; (2) mounting that is an integral part of oestrus behaviour and is regulated by the same hormones as oestrus ; and (3) mounting that is induced by treatment with large (often pharmacological) quantities of hormones that are normally present in only small quantities in the female, either oestrogens or androgens. However, not all species display all these types of mounting. For example, female hamsters, dogs, and rhesus monkeys, unlike female rats or guinea pigs, do not show augmentation of mounting when treated with testosterone unless they have also been exposed to testosterone early in development. Female guinea pigs mount when in oestrus, when ovariectomized and treated sequentially with oestradiol and progesterone, and with testosterone. Anoestrus and ovariectomized, untreated females mount, but only very infrequently. This low baseline frequency that is characteristic of the spayed guinea pig can be altered by prenatal exposure to androgens or oestrogens. Clearly then, there are distinct subsets or classes of mounting behaviour that can be distinguished not only by their unique susceptibilities to hormonal influences in adulthood, but also by their hormonal regulation during early developmental stages.

Why then do females of some species only mount in some endocrine conditions ? The observation that many neural sites contribute to the regulation of mounting behaviour in males leads to the possibility that the different classes of mounting are regulated by different neuronal populations, each of which has genetically encoded sensitivity to different steroids. This is supported by evidence in the guinea pig that lesions of the hypothalamus can disrupt that mounting which is normally induced by the hormones of oestrus (oestradiol and progesterone), but not that which is androgen-induced. Mounting in the gonadectomized female and male guinea pig may be regulated by those neurons which are also androgen sensitive as well as by a third neuronal population sensitive to prenatal androgens or oestrogens. Species differences in the ability to display these types of mounting may be explained by differential sensitivities to hormones (i.e. the adult female guinea pig may be more sensitive to testosterone than the female hamster) or to the amount of steroid exposure early in development (i.e. the female rat, which mounts following ovariectomy, may be exposed to more steroid in utero than the female guinea pig).

Alternatively, mounting in each of the endocrine conditions, or mounting directed to different partners may involve unique contributions from several brain areas. We have pointed out that the hypothalamus, limbic structures, and cortex can all regulate the display of mounting. The degree to which these, and possibly other, structures are involved in the regulation of androgen-induced mounting and oestradiol progesterone-induced mounting may be quite different. Mounting of male and female partners too, is certain to involve different cues, and the types of cortical processing that occur with each partner may be quite different. This would result in unique involvement of the cortex with each sex partner.

Finally, these two hypotheses need not be mutually exclusive. We have pointed out that many areas of the brain concentrate steroid hormones. Thus, both differential hormonal sensitivities and contributions from various neural structures may result in the display of mounting behaviour by females in some environmental conditions and not in others.

By dividing female mounting behaviour into three classes based on their endocrine regulation, it becomes apparent that, despite similarities in appearance, these are distinct types of mounting which are not all homologies of the sexual mounting displayed by males. For example, Feder and Goy (1983) argued that the mounting of female guinea pigs that was displayed under the influence of ovarian hormones was a component of normal proceptive behaviour and not an indicator of masculine behaviour, and, such mounting does not serve as an index of the bisexual potential of the female guinea pig. Hines and Goy (1985) held a somewhat similar view for guinea pigs, suggesting in addition that only the mounting activity displayed in the absence of gonadal hormones or the mounting activated by exogenous testosterone should be regarded as masculine homologues. Similar considerations led Thornton (1986) to propose the terminology of "true hetero-typical" and "quasi-heterotypical" for mounting associated with masculinization and mounting that was expressed by females independently of known masculinizing processes respectively.

This discussion would be incomplete if we did not address the observation of many investigators that female heterotypical behaviour subserves a variety of functions. This seems especially apparent in studies comparing isosexual and heterosexual mounting. Dagg (1984) has documented female-female mounting in 71 mammalian species with female-male mounting occurring in 43 of them. Furthermore, mounting was seen in several different contexts such as play, aggression and sexual encounters (Dagg, 1984). Obviously then, to treat mounting by females globally and ascribe a single function to it fails to consider the many variables that may influence this behaviour. Several functions have been proposed for female mounting including the assertion of dominance, social bonding, and stimulation of a partner's sexual response.

Mounting that occurs in the anoestrus or ovariectomized female may be related to the expression of dominance. In the juvenile female rhesus, the frequency of mounting was greatest in high ranking females and lowest in low ranking females. Alternatively, Parker and Pearson (1976) have proposed that his type of mounting may be reciprocal altruism. That is, it serves to attract a mate to an individual genetically related to the mounter who is able to mate with the male. This suggestion is plausible, but only in situations where there is one male that does most of the mating and there are many inter-related females. These social groupings occur in ungulates and in some primate species where female-female mounting has been observed.

Mounting by oestrus females may also serve several functions. Parker and Pearson (1976) have suggested that female-female mounting is an alternative to proceptive solicitation, and, as such, serves to attract the dominant male and ensure mating. However, they offer no rationale for mounting of male partners. This has been addressed by Ford and Beach (1952) and Morris (1955) who suggest that in rats and dogs female mounting may serve to arouse a sluggish male who is slow to mate. In this instance, Morris suggests that mounting is a resolution of the conflicting tendencies of fleeing, attacking, and mating.

Neither of the hypotheses advanced by Parker and Pearson (1976) or Ford and Beach (1952) and Morris (1955) adequately explain why female-female mounting occurs in the absence of males, the mounting of both male and female partners by post-conception japanese macaques, or the observation in stumptail macaques that the mounting female was always lactating. In these instances, mounting may serve to strengthen social bonding or serve some other, as of yet, undefined function.

There are still several questions about female mounting that have not been addressed by investigators. For example, only females of some species mount to express dominance, to stimulate sexual responses of males, or for social bonding. Why aren't all of these types of mounting seen in all species ? Why do females of some species only mount other females while, in other species, females mount partners of either sex. These questions would imply that perhaps the conditions for mounting, both endocrine and social, are not universal.

Most past studies of mounting behaviour have been limited to examining a particular variable that may influence mounting in a particular set of stimulus conditions without examining mounting that occurs in other situations. Only now is the study of mounting advancing so that an accurate comparative picture can be anticipated. Such comparative studies are needed on a broad scale to include non-social as well as social species and non-mammalian vertebrates for what they can contribute to an understanding of phylogeny. We ought to mention in this regard the remarkable and thought provoking studies of heterotypical behaviour of a parthenogenetic species of lizard by David Crews (Moore & al., 1985 ; Grassman & Crews, 1986).

REFERENCES

Akers, J.S. and Conaway, C.H. (1979). Female homosexual behavior in Macaca mulatta. Archives of Sexual Behavior, 8, 63-80.

Arendash, G.W. and Gorski, R.A. (1983). Effects of discrete lesions of the sexually dimorphic nucleus of the preoptic area or other medial preoptic regions on the sexual behavior of male rats. Brain Research Bulletin, 10, 147-154.

Ayoub, D.M., Greenough, W.T. and Juraska, J.M. (1983). Sex differences in dendritic structure in the preoptic area of the juvenile macaque monkey brain. Science, 219, 197-198.

Ball, J. (1937). The effect of male hormone on the sex behavior of female rats. Psychological Bulletin, 34, 725.

Ball, J. (1940). The effect of testosterone on the sex behavior of female rats. Journal of Comparative Psychology, 29, 151-165.

Baum, M.J. (1976). Effects of testosterone propionate administered perinatally on sexual behavior of female ferrets. Journal of Comparative and Physiological Psychology, 90, 399-410.

Baum, M.J., Gallagher, C.A., Martin, J.T. and Damassa, D.A. (1982a). Effects of testosterone, dihydrotestosterone, or estradiol administered neonatally on sexual behavior of female ferrets. Endocrinology, 111, 773-780.

Baum, M.J., Tobet, S.A., Starr, M.S. and Bradshaw, W.G. (1982b). Implantation of dihydrotestosterone propionate into the lateral septum or medial amygdala facilitates copulation in castrated male rats given estradiol systemically. Hormones and Behavior, 16, 208-223.

Beach, F.A. (1940). Effects of cortical lesions upon the copulatory behavior of male rats. Journal of Comparative Psychology, 29, 193-244.

Beach, F.A. (1942). Male and female mating behavior in prepuberally castrated female rats treated with androgens. Endocrinology, 31, 673-678.

Beach, F.A. (1943). Effects of injury to the cerebral cortex upon the display of masculine and feminine behavior by female rats. Journal of Comparative Psychology, 36, 169-198.

Beach, F.A. (1968). Factors involved in the control of mounting behavior by female mammals. In : Reproduction and Sexual Behavior, Diamond M. (Ed.), Indiana University Press, pp. 31-81.

Beach, F.A. and Rasquin, P. (1942). Masculine copulatory behavior in intact and castrated female rats. Endocrinology, 31, 393-409.

Beach, F.A., Rogers, C.M. and LeBoeuf, B.J. (1968). Coital behavior in dogs : effects of estrogen on mounting by females. Journal of Comparative and Physiological Psychology, 66, 296-307.

Beach, F.A., Kuehn, R.E., Sprague, R.A. and Anisko, J.J. (1972). Coital behavior in dogs. XI. Effects of androgenic stimulation during development on masculine mating responses in females. Hormones and Behavior, 3, 143-168.

Bleier, R., Byne, W. and Siggelkow, I. (1982). Cytoarchitectonic sexual dimorphisms of the medial preoptic and anterior hypothalamic areas in guinea pig, rat, hamster, and mouse. Journal of Comparative Neurology, 212, 118-130.

Bonneau, M., Ahdieh, A.B., Thornton, J.E. and Feder, H.H. (1987). Cytosol androgen receptors in guinea pig brain and pituitary. Brain Research, 413, 104-110.

Brookhart, J.M. and Dey, F.L. (1941). Reduction of sexual behavior in male guinea pigs by hypothalamic lesions. American Journal of Physiology, 133, 551-554.

Butera, P. and Czaja, J. (1985). Effects of intracranial dihydro-testosterone on the reproductive physiology and behavior of male guinea pigs. Abstract. Conference on Reproductive Behavior.

Byne, W. and Bleier, R. (1987). Medial preoptic sexual dimorphisms in the guinea pig. I. An investigation of their hormonal dependance. Journal of Neurosciences, 7, 2688-2696.

Carter, C.S., Clemens, L.G. and Hoekema, D.J. (1972). Neonatal androgen and adult sexual behavior in the golden hamster. Physiology and Behavior, 9, 89-95.

Chevalier-Skolnikoff, S. (1976). Homosexual behavior in a laboratory group of stumptail monkeys (Macaca arctoides) : forms, contexts and possible social functions. Archives of Sexual Behavior, 5, 511-527.

Christensen, L.W. and Clemens, L.G. (1974). Intrahypothalamic implants of testosterone or estradiol and resumption of masculine sexual behavior in long-term castrated male rats. Endocrinology, 95, 984-990.

Christensen, L.W. and Gorski, R.A. (1978). Independent masculinization of neuroendocrine systems by intracerebral implants of testosterone or estradiol in the neonatal female rat. Brain Research, 146, 325-340.

Clemens, L.G. and Coniglio, L. (1971). Influence of prenatal litter composition on mounting behavior of female rats. American Zoologist, 11, 617-618.

Clemens, L.G., Gladue, B.A. and Coniglio, L.P. (1978). Prenatal endogenous androgenic influences on masculine sexual behavior and genital morphology in male and female rats. Hormones and Behavior, 10, 40-53.

Commins, D. and Yahr, P. (1984). Adult testosterone levels influence the morphology of a sexually dimorphic area in the mongolian gerbil brain. Journal of Comparative Neurology, 224, 132-140.

Dagg, A.I. (1984). Homosexual behaviour and female-male mounting in mammals : a first survey. Review of Mammalogy, 14, 155-185.

Davidson, J.M. (1966). Activation of the male rat's sexual behavior by intracerebral implantation of androgen. Endocrinology, 79, 783-794.

De Jonge, F.H., Burger, J. and Van de Poll, N.E. (1986). Variable mounting levels in the female rat : the influence of experience and acute effects of testosterone. Behavioral and Brain Research, 20, 39-46.

De Olmos, J.S. (1972). The amygdaloid projection field in the rat as studied with the cupric silver method. In : Advances in Behavioral Biology Vol. 2, The Neurobiology of the Amygdala, Eleftheriou B.E. (Ed.), Plenum Press, New York, pp. 145-204.

De Waal, F.B.M. (1989). La réconciliation chez les primates. La Recherche, 20, 588-597.

Dicks, D., Myers, R.E. and Kling, A. (1969). Uncus and amygdala lesions : effects on social behavior in the free ranging rhesus monkey. Science, 165, 69-71.

Döhler, K.D., Coquelin, A., Davis, F., Hines, M., Shryne, J.E. and Gorski, R.A. (1982). Differentiation of the sexually dimorphic nucleus in the preoptic area of the rat brain is determined by the perinatal hormone environment. Neurosciences Letters, 33, 295-298.

Döhler, K.D., Coquelin, A., Davis, F., Hines, M., Shryne, J.E. and Gorski, R.A. (1984). Pre- and postnatal influence of testosterone propionate and diethylstilbestrol on differentiation of the sexually dimorphic nucleus of the preoptic area in male and female rats. Brain Research, 302, 291-295.

Ehrhardt, A.A., Meyer-Bahlburg, H.F.L., Rosen, L.R., Feldman, J.F., Veridiano, N.P., Zimmerman, I. and McEwen, B.S. (1985). Sexual orientation after prenatal exposure to exogenous estrogen. Archives of Sexual Behavior, 14, 57-75.

Feder, H.H. and Goy, R.W. (1983). Effects of neonatal estrogen treatment of female guinea pigs on mounting behavior in adulthood. Hormones and Behavior, 17, 284-291.

Feder, H.H., Resko, J.A. and Goy, R.W. (1968). Progesterone concentrations in the arterial plasma of guinea pigs during the oestrous cycle. Journal of Endocrinology, 40, 505-513.

Fedigan, L.M. and Gouzoules, H. (1978). The consort relationship in a troop of japanese monkeys. In : Recent Advances in Primatology Vol. 1, Chivers D.J. and Herbert J. (Eds.), Academic Press, New York, pp. 489-495.

Ford, C.S. and Beach, F.A. (1952). Patterns of Sexual Behavior, Harper, New York.

Gerall, A.A. (1966). Hormonal factors influencing masculine behavior of female guinea pigs. Journal of Comparative and Physiological Psychology, 62, 365-369.

Giantonio, G.W., Lund, N.L. and Gerall, A.A. (1970). Effect of diencephalic and rhinencephalic lesions on the male rat's sexual behavior. Journal of Comparative and Physiological Psychology, 73, 38-46.

Ginton, A. and Merari, A. (1977). Long range effects of MPOA lesion on mating behavior in the male rat. Brain Research, 120, 158-163.

Gladue, B.A. (1984). Dihydrotestosterone stimulates mounting behavior but not lordosis in female rats. Physiology and Behavior, 33, 49-53.

Goldfoot, D.A. (1979). Sex-specific, behavior-specific actions of dihydrotestosterone : activation of aggression, but not mounting in ovariectomized guinea pigs. Hormones and Behavior, 13, 241-255.

Goldfoot, D.A. and Van der Werff ten Bosch (1975). Mounting behavior of female guinea pigs after prenatal and adult administration of the propionates of testosterone, dihydrotestosterone and androstanediol. Hormones and Behavior, 6, 139-148.

Golub, M.S., Hayes, L., Prahalada, S. and Hendrickx, A.G. (1983). Behavioral tests in monkey infants exposed embryonically to an oral contraceptive. Neurobehavioral and Toxicological Teratology, 5, 301-304.

Gorski, R.A., Gordon, J.H., Shryne, J.E. and Southam, A.M. (1978). Evidence for a morphological sex difference within the medial preoptic area of the rat Brain. Brain Research, 148, 333-346.

Gorski, R.A., Harlan, R.E., Jacobsen, C.D., Shryne, J.E. and Southam, A.M. (1980). Evidence for the existence of a sexually dimorphic nucleus in the preoptic area of the rat. Journal of Comparative Neurology, 193, 529-539.

Gouzoules, H. and Goy, R.W. (1983). Physiological and social influences on mounting behavior of troop living female monkeys (Macaca fuscata). American Journal of Primatology, 5, 39-49.

Goy, R.W. (1978). Development of play and mounting behavior in female rhesus virilized prenatally with esters of testosterone or dihydrotestosterone. In : Recent Advances in Primatology Vol. 1, Chivers D. and Herbert J. (Eds.), Academic Press, New York, pp. 449-462.

Goy, R.W. (1981). Differentiation of male social traits in female rhesus macaques by prenatal treatment with androgens : variation in type of androgen, duration, and timing of treatment. In : Fetal Endocrinology, Novy M.J. and Resko J.A. (Eds.), Academic Press, New York, pp. 319-339.

Goy, R.W. and Jakway, J.S. (1959). The inheritance of patterns of sexual behavior in female guinea pigs. Animal Behaviour, 7, 142-149.

Goy, R.W. and Kemnitz, J.W. (1983). Early, persistant, and delayed effects of virilizing substances delivered transplacentally to female rhesus fetuses. In : Application of Behavioral Pharmacology in Toxicology, Zbinden G. (Ed.), Raven Press, New York, pp. 303-314.

Goy, R.W. and Phoenix, C.H. (1963). Hypothalamic regulation of female sexual behavior : establishment of behavioural oestrus in spayed guinea-pigs following hypothalamic lesions. Journal of Reproduction and Fertility, 5, 23-40.

Goy, R.W. and Young, W.C. (1957). Strain differences in the behavioural responses of female guinea pigs to alpha-estradiol benzoate and progesterone. Behaviour, 10, 340-354.

Goy, R.W., Bridson, W.E. and Young, W.C. (1964). The period of maximal susceptibility of the prenatal female guinea pig to masculinizing actions of testosterone propionate. Journal of Comparative and Physiological Psychology, 57, 166-174.

Goy, R.W., Bercovitch, F.B. and McBrair, M.C. (1988). Behavioral masculinization is independent of genital masculinization in prenatally androgenized female rhesus macaques. Hormones and Behavior, 22, 552-571.

Grassmann, M. and Crews, D. (1986). Progesterone induction of pseudocopulatory behavior and stimulus-response complementarity in an all-female lizard species. Hormones and Behavior, 20, 327-335.

Gray, P. and Brooks, P.J. (1984). Effect of lesion location within the medial preoptic-anterior hypothalamic continuum on maternal and male sexual behavior in female rats. Behavioral Neurosciences, 98, 703-711.

Greenough, W.T., Carter, C.S., Steerman, C. and DeVoogd, T.J. (1977). Sex differences in dendritic patterns in hamster preoptic area. Brain Research, 126, 63-72.

Hart, B.L. and Ladewig, J. (1979). Effects of medial preoptic-anterior hypothalamic lesions on development of sociosexual behavior in dogs. Journal of Comparative and Physiological Psychology, 93, 566-573.

Heimer, L. and Larsson, K. (1967). Impairment of mating behavior in male rats following lesions in the preoptic-anterior hypothalamic continuum. Brain Research, 3, 248-263.

Hines, M. (1982). Prenatal gonadal hormones and sex differences in human behavior. Psychological Bulletin, 92, 56-80.

Hines, M. and Goy, R.W. (1985). Estrogens before birth and development of sex-related reproductive traits in the female guinea pig. Hormones and Behavior, 19, 331-347.

Hines, M., Davis, F.C., Coquelin, A., Goy, R.W. and Gorski, R.A. (1985). Sexually dimorphic regions in the medial preoptic area and the bed nucleus of the stria terminalis of the guinea pig brain : a description and investigation of their relationship to gonadal steroids in adulthood. Journal of Neurosciences, 5, 40-47.

Hines, M., Alsum, P., Roy, M., Gorski, R.A. and Goy, R.W. (1987). Estrogenic contributions to sexual differentiation in the female guinea pig : influences of diethylstilbestrol and tamoxifen on neural, behavioral and ovarian development. Hormones and Behavior, 21, 402-417.

Hsu, C.H. and Carter, C.S. (1986). Social isolation inhibits male-like sexual behavior in female hamsters. Behavioral and Neural Biology, 46, 242-247.

Johnston, P. and Davidson, J.M. (1972). Intracerebral androgens and sexual behavior in the male rat. Hormones and Behavior, 3, 345-357.

Kling, A. (1968). Effects of amygdalectomy and testosterone on sexual behavior of male juvenile macaques. Journal of Comparative and Physiological Psychology, 65, 466-471.

Klüver, H. and Bucy, P.C. (1937). "Psychic Blindness" and other symptoms following bilateral temporal lobectomy in rhesus monkeys. American Journal of Physiology, 119, 352-353.

Krieger, M.S., Morrell, J.I. and Pfaff, D.W. (1976). Auto-
    radiographic localization of estradiol-containing cells in the
    female hamster brain. Neuroendocrinology, 22, 193-205.
Larsson, K. (1962). Mating behavior in male rats after cerebral
    cortical ablation. I. Effects of lesions in the dorsolateral and
    median cortex. Journal of Experimental Zoology, 151, 167-176.
Larsson, K. (1964). Mating behavior in male rats after cerebral
    cortical ablation. II. Effects of lesions in the frontal lobes
    compared to lesions in the posterior half of the hemispheres.
    Journal of Experimental Zoology, 155, 203-214.
Lehman, M.N., Winans, S.S. and Powers, J.B. (1980). Medial nucleus
    of the amygdala mediates chemosensory control of male
    sexual behavior. Science, 210, 557-560.
Lisk, R.D. (1968). Copulatory activity of the male rat following
    placement of preoptic-anterior hypothalamic lesions. Experi-
    mental Brain Research, 5, 306-313.
Lisk, R.D. and Bezier, J.L. (1980). Intrahypothalamic hormone
    implantation and activation of sexual behavior in the male
    hamster. Neuroendocrinology, 30, 220-227.
Lisk, R.D. and Greenwald, D.P. (1983). Central plus peripheral
    stimulation by androgen is necessary for complete restoration
    of copulatory behavior in male hamsters. Neuroendocrinology,
    36, 211-217.
MacLusky, N.J., Chaptal, C. and McEwen, B.S. (1979a). The
    development of estrogen-receptor systems in the rat brain :
    postnatal development. Brain Research, 178, 143-160.
MacLusky, N.J., Lieberburg, I. and McEwen, B.S. (1979b). The
    development of estrogen-receptor systems in the rat brain :
    perinatal development. Brain Research, 178, 129-142.
McEwen, B.S. (1980). Binding and metabolism of sex steroids by
    the hypothalamic-pituitary unit : physiological implications.
    Annual Review of Physiology, 42, 97-110.
Michael, R.P., Wilson, M.I. and Zumpe, D. (1974). The bisexual
    behavior of female rhesus monkeys. In : Sex Differences in
    Behavior, Friedman R.C., Richart R.M. and Van de Wiele
    R.L. (Eds.), Wiley, New York, pp. 399-412.
Money, J. and Schwartz, M. (1977). Dating, romantic and non ro-
    mantic friendships and sexuality in 17 early-treated adreno-
    genital females, aged 16-25. In : Congenital Adrenal Hyper-
    plasia. Lee P.A., Plotnick L.P., Kowartki A.A. and Migeon
    C.J. (Eds.), University Park Press, Baltimore, pp. 419-431.
Money, J., Ehrhardt, A.A. and Masica, D.N. (1967). Fetal femini-
    zation indiced by androgen insensitivity in the testicular
    feminization syndrome : effects on marriage and materna-
    lism. Johns Hopkins Medical Journal, 123, 104-114.
Moore, M.C., Whittier, J.M., Billy, A.J. and Crews, D. (1985).
    Male-like behavior in an all-female lizard : relationship to
    ovarian cycle. Animal Behavior, 33, 284-289.
Morris, D. (1955). The causation of pseudofemale and pseudomale
    behaviour : a further comment. Behaviour, 8, 46-56.

Olsen, K.L. (1979). Induction of male mounting in androgen-insensitive (tfm) and normal (King-Holtzman) male rats : effect of testosterone propionate, estradiol benzoate, and dihydrotestosterone. Hormones and Behavior, 13, 66-84.

Olsen, K.L. and Whalen, R.E. (1981). Hormonal control of the development of sexual behavior in androgen-insensitive (tfm) rats. Physiology and Behavior, 27, 883-886.

Parker, G.A. and Pearson, R.G. (1976). A possible origin and adaptive significance of the mounting behaviour shown by some female mammals in oestrus. Journal of Natural History, 10, 241-245.

Paup, D.C., Coniglio, L.P. and Clemens, L.G. (1972). Masculinization of the female golden hamster by neonatal treatment with androgen or estrogen. Hormones and Behavior, 3, 123-131.

Paxinos, G. (1976). Interruption of septal connections : effects on drinking, irrability, and copulation. Physiology and Behavior, 17, 81-88.

Pfaff, D.W. (1968). Autoradiographic localization of radioactivity in rat brain after injection of tritiated sex hormones. Science, 161, 1355-1356.

Pfaff, D.W. (1970). Nature of sex hormone effects on rat sex behavior : specificity of effects and individual patterns of response. Journal of Comparative and Physiological Psychology, 73, 349-358.

Pfaff, D.W. and Zigmond, R.E. (1971). Neonatal androgen effects on sexual and non-sexual behavior of adult rats tested under various hormone regimens. Neuroendocrinology, 7, 129-145.

Phoenix, C.H. (1961). Hypothalamic regulation of sexual behavior in male guinea pigs. Journal of Comparative and Physiological Psychology, 54, 72-77.

Phoenix, C.H., Goy, R.W., Gerall, A.A. and Young, W.C. (1959). Organizing action of prenatally administered testosterone propionate on the tissues mediating mating behavior in the female guinea pig. Endocrinology, 65, 369-382.

Pomerantz, S.M., Fox, T.O., Sholl, S.A., Vito, C.C. and Goy, R.W. (1982). Androgen and estrogen receptors in the fetal rhesus monkey brain and anterior pituitary. Endocrinology, 116, 83-89.

Pomerantz, S.M., Roy, M.M., Thornton, J.E. and Goy, R.W. (1985). Expression of adult female patterns of sexual behavior by male, female and pseudohermaphroditic female rhesus monkeys. Biology of Reproduction, 33, 878-889.

Pomerantz, S.M., Goy, R.W. and Roy, M.M. (1986). Expression of male-typical behavior in adult female pseudohermaphroditic rhesus : comparisons with normal males and neonatally gonadectomized males and females. Hormones and Behavior, 20, 483-500.

Pomerantz, S.M., Roy, M.M. and Goy, R.W. (1988). Social and hormonal influences on behavior of adult male, female and pseudohermaphroditic rhesus monkeys. Hormones and Behavior, 22, 219-230.

Pope, N.S., Wilson, M.E. and Gordon, T.P. (1987). The effect of season on the induction of sexual behavior by estradiol in female rhesus monkeys. Biology of Reproduction, 36, 1047-1054.

Rainbow, T.C., Parsons, B. and McEwen, B.S. (1982). Sex differences in rat brain oestrogen and progestin receptors. Nature, 300, 648-649.

Raisman, G. and Field, P.M. (1973). Sexual dimorphism in the neuropil of the preoptic area of the rat and its dependence on neonatal androgen. Brain Research, 54, 1-29.

Rees, H.D. and Michael, R.P. (1982). Brain cells of the male rhesus monkey accumulate 3H-testosterone or it's metabolites. Journal of Comparative Neurology, 206, 273-277.

Roy, M.M. and Goy, R.W. (1988). Sex differences in the inhibition by ATD of testosterone-activated mounting behavior in guinea pigs. Hormones and Behavior, 22, 315-323.

Sachs, B.D. and Thomas, D.A. (1985). Differential effects of perinatal androgen treatment on sexually dimorphic characteristics in rats. Physiology and Behavior, 34, 735-742.

Sachs, B.D., Pollak, E.I., Krieger, M.S. and Barfield, R.J. (1973). Sexual behavior : normal male patterning in androgenized female rats. Science, 181, 770-772.

Shapiro, B.H., Levine, D.C. and Adler, N.T. (1980). The testicular feminized rat : a naturally occuring model of androgen independent brain masculinization. Science, 209, 418-420.

Singer, J.J. (1968). Hypothalamic control of male and female sexual behavior in female rats. Journal of Comparative and Physiological Psychology, 66, 738-742.

Slimp, J.C., Hart, B.L. and Goy, R.W. (1978). Heterosexual, autosexual and social behavior of adult male rhesus with medial preoptic-anterior hypothalamic lesions. Brain Research, 142, 105-122.

Södersten, P. (1972). Mounting behavior in the female rat during the estrous cycle, after ovariectomy, and after estrogen or testosterone administration. Hormones and Behavior, 3, 307-320.

Södersten, P. (1973). Increased mounting behavior in the female rat following a single injection of testosterone propionate. Hormones and Behavior, 4, 1-17.

Spies, H.G., Norman, R.L., Clifton, D.K., Ochsner, A.J., Jensen, J.N. and Phoenix, C.H. (1976). Effects of bilateral amygdaloid lesions on gonadal and pituitary hormones in serum and on sexual behavior in female rhesus monkeys. Physiology and Behavior, 17, 985-992.

Stumpf, W.E. (1970). Estrogen-neurons and estrogen-neuron systems in the periventricular brain. American Journal of Anatomy, 129, 207-218.

Swanson, H.H. (1971). Determination of the sex role in hamsters by the action of sex hormones in infancy. In : Influences of Hormones on the Nervous System, Ford D.H. (Ed.), Proceedings of the International Society of Psychoneuroendocrinology, pp. 424-440.

Swanson, L.W. (1976). An autoradiographic study of the efferent connections of the preoptic region in the rat. Journal of Comparative Neurology, 167, 227-256.

Tiefer, L. (1970). Gonadal hormones and mating behavior in the adult golden hamster. Hormones and Behavior, 1, 189-202.

Thornton, J.E. (1986). Heterotypical sexual behavior : implications from variations. In : Reproduction : A Behavioral and Neuro-endocrine Perspective, Komisaruk B.R., Siegel H.I., Cheng M.F. and Feder H.H. (Eds.), Annals of the New York Academy of Sciences, pp. 362-370.

Tobet, S.A., Zahniser, D.J. and Baum, M.J. (1986). Differentiation in male ferrets of a sexually dimorphic nucleus of the preoptic/anterior hypothalamic area requires prenatal estrogen. Neuroendocrinology, 44, 299-308.

Vito, C.C. and Fox, T.O. (1982). Androgen and estrogen receptors in embryonic and neonatal rat brain. Developmental Brain Research, 2, 97-110.

Von Lawick-Goodall, J. (1968). The behaviour of free-living chimpanzees in Gombe Stream Reserve. Animal Behaviour Monographs, 5, 85-198.

Wallen, K. (1978). Genetic and hormonal influences on the bisexuality of females from two inbred strains of guinea pig. Doctoral Dissertation, University of Wisconsin-Madison.

Ward, I.L. (1969). Differential effect of pre- and postnatal androgen on the sexual behavior of intact and spayed female rats. Hormones and Behavior, 1, 25-36.

Warembourg, M. (1977). Radioautographic localization of estrogen-containing cells in the brain and pituitary of the guinea pig. Brain Research, 123, 357-362.

Whalen, R.E., Edwards, D.A., Luttge, W.G. and Robertson, R.T. (1969). Early androgen treatment and male sexual behavior in female rats. Physiology and Behavior, 4, 33-39.

Yerkes, R.M. (1939). Social dominance and sexual status in the chimpanzee. Quarterly Review of Biology, 14, 115-136.

Young, W.C. and Rundlett, B. (1939). The hormonal induction of homosexual behavior in the spayed female guinea pig. Psychosomatic Medicine, 1, 449-460.

Young, W.C., Dempsey, E.W. and Hyers, H.I. (1935). Cyclic reproductive behavior in the female guinea pig. Journal of Comparative Psychology, 19, 313-335.

Young, W.C., Dempsey, E.W., Hagquist, C.W. and Boling, J.L. (1939). Sexual behavior and sexual receptivity in the female guinea pig. Journal of Comparative Psychology, 27, 49-68.

5 Heterotypical Sexual Behaviour in Male Mammals : The Rat as an Experimental Model

Claude Aron, Danielle Chateau, Christiane Schaeffer and Jacqueline Roos *

In the first chapter of <u>Three essays on the theory of sexuality</u>, Freud, while discussing the mechanisms of homosexuality in the human, referred to the hypothesis formulated by Krafft-Ebing (1931) of the existence of a "male and female brain" within the brain of both male and female individuals, that could account for the capacity of individuals of both sexes to express either heterosexual and homosexual behaviour. This was really a remarkably foresighted view in the light of present knowledge on the neuroendocrine mechanisms governing sexual behaviour in mammals. Moreover it has been known for some years that homosexuality spontaneously occurs in females of a large number of subhuman mammalian species where females mount other animals in a manner resembling the copulatory behaviour of males of the same species. Although less frequently, female copulatory behaviour has also been shown to occur in males.

It is essential to define homosexuality in non-human species since no information is available concerning the sexual motivation of either partner of a homosexual pair of animals. Does mounting of a male dog by another male correspond to a state of aspecific non-sexual arousal or, in Freudian terms, to occasional homosexuality ? Which is the homosexual partner in a pair of rats in which a female mounts another female ? We believe that only male or female individuals that display the copulatory pattern typical of the opposite sex i.e. a male which display a lordotic response to male mounts or a female which exhibits the characteristic male pattern of pelvic thrusting and intromission while mounting another female .may be considered as homosexual during sexual interactions with a partner of the same sex. The term heterotypical sexual behaviour must thus be used to designate a kind of behaviour that is in disaccord with the genetic sex. In contrast,

---

* Institut d'Histologie, Faculté de Médecine, 4, rue Kirschleger, 67085 Strasbourg Cedex, France.

term homotypical behaviour will refer to the typical male or
female copulatory pattern of each sex. Thus both males and females of some mammalian species possess a nervous and muscular
organization capable of mediating heterotypical behaviour. It
should, however, be recognized that heterotypical sexual behaviour
represents a rather infrequent form of sexuality at least in
subhuman male mammals. This is probably due to the fact that
development of homotypical sexual behaviour depends on steroid
hormones which control the organization of neural systems at
early critical periods in the male and the subsequent activation in
adulthood of such neural pathways that mediate sex-related
behaviour in both sexes by sex steroids (testosterone and/or
dihydrotestosterone in the male and oestradiol and progesterone in
the female). Thus the spontaneous display of heterotypical
behaviour in the male of lower mammals may result from impairment either of organizational or activational processes or both. In
addition it is necessary to examine whether olfactory signals,
which fundamentally influence homotypical sexual interactions in
mammals, also modulate the activational effects of steroid hormones for the display of heterotypical sexual behaviour.

   This chapter has three objectives. The first is to show that
despite the organizational effects of androgens during the
perinatal period, the males of different mammalian species have
the capacity to display feminine behaviour in adulthood. The
second objective is to analyse the neuroendocrine and olfactory
mechanisms involved in the display of feminine behaviour in the
male rat, and the last is to evaluate in the light of
physiopathological and experimental data the respective
involvement of environmental and organizational factors in
determining sexual orientation in human beings.

ORGANIZATION OF MALE SEXUAL BEHAVIOUR

Morphological and Enzymological Changes Caused by Androgens in
the Brain During Foetal and Perinatal Periods

   The display of sexual behaviour in the male is dependent upon
the functional integrity of a nervous hypothalamic structure called
the preoptic area (POA) while in the female it depends on the
ventromedial nucleus (VMN) situated in the mediobasal hypothalamus (MBH). In both the male and female, those structures
develop, and may be activated heterotypically in adulthood by
hormones of the opposite sex. The mechanisms by which sex
steroids organize neural structures and neural pathways responsible
for homotypical sexual behaviour in the male have thus to be
defined. It has been known for several years that, if androgens are
present during the perinatal period, neural and behavioural
development is male-oriented. In the male rat, the testis produces

androgens from day 15.5 of pregnancy (which lasts 21-22 days) until day 4 after birth (Habert & Picon, 1984 ; Csernus, 1986). Two peaks of androgen secretion occur, the first at the end of pregnancy (Weisz & Ward, 1980) and the second immediately after birth (Corbier & al., 1978 ; Gogan & al., 1981 ; Roffi & al., 1987). In the human the interstitial gland of the testis develops during foetal life and produces androgens from the 8th week in utero onwards (Siiteri & Wilson, 1974). Other reports indicate that androgen secretion peaks at week 7-11 and declines from the 17th to the 21st week (Reyes & al., 1974). In the baboon, testes at foetal age 100-107 days (comparable to human foetus testes of mid-gestation) also secrete testosterone, using adrenocortical dehydroepiandrosterone as a substrate (Redmond & al., 1986). We may thus assume that data on the hormonal milieu at certain stages of pregnancy are compatible with the hypothesis that androgens may affect the central nervous system.

Two hypothalamic areas, the POA and the VMN, are especially representative of such organizational effects. For example, in the rat, large sexual difference in the volume of a densely staining component of the POA, called sexually dimorphic nucleus (SDN-POA) (Gorski & al., 1980), is observed shortly after birth. Steroid manipulations modify the development of this nucleus (Gorski & al., 1978) which is much larger in males than in females and the cells of which are larger and more numerous in males. A dimorphic organization of the enkephalinergic fibres in the periventricular preoptic nucleus has been also observed by Wiegand & al. (1986). Plexuses of fibres were shown to be more dense in females than in males. Neonatal androgenization of females completely suppressed this feminine pattern in the periventricular preoptic nucleus (Wiegand & al., 1986).

Recently, Matsumoto and Arai (1986) reported the results of ultrastructural and experimental observations in the rat suggesting that the presence of androgens in the neonate was responsible for the development of sexual dimorphism in the synaptic organization of the ventrolateral part of the VMN (VL-VMN). In normal males, the number of shaft and spine synapses was greater in the VL-VMN than in the dorsomedial part of the VMN (DM-VMN). In females, however, no significant difference in the number of shaft and spine synapses was noted between the DM-VMN and the VL-VMN. Moreover the number of synapses in VL-VMN was greater in males than in females. Castration of males on post-natal day 1 reduced the number of synapses in VL-VMN to a level comparable to that of normal females. The well known dimorphic pattern of steroid concentrating neurones in the VMN (Pfaff & Keiner, 1973 ; Stumpf, 1970) which are more abundant in the VL-VMN, is consistent with the sexually dimorphic organization of this hypothalamic nucleus.

The metabolic pathways which are involved in the action of testosterone at the level of the central nervous system have been known for several years. It has been demonstrated that a mechanism of aromatization results in the transformation of testosterone and androstenedione into oestrogen (Naftolin & al., 1975). Limbic and hypothalamic structures appeared to be the main sites of this conversion of androgens into oestrogens during foetal and adult life. It has been shown that androstenedione can be converted into oestrone and oestradiol by the hypothalamus and the median eminence of human foetuses at 10 and 22 weeks (Naftolin & al., 1972). Conversion of androstenedione to oestrone has also been observed in neural tissue from foetal and neonatal rats (Reddy & al., 1974). Aromatase activity is known to be present during the period corresponding to the endocrine activity of the testis at the end of pregnancy and early after birth in the rat since the hypothalamus is capable of converting testosterone into oestradiol between days 17 and 20 of pregnancy (Weisz & al., 1982) and on days 1-4 after birth (George & Ojeda, 1982). In fact, it is during this critical time period that the organizational effects of androgens occur.

Masculinization and Defeminization Effect of Androgens

Masculinization

The use of aromatase inhibitors or antiandrogenic drugs during the perinatal period of the rat has produced a clear understanding of the chronology of the neuroendocrine events responsible for behavioural masculinization and/or defeminization of genetically male subjects. The masculinization effects of androgens are produced early during pregnancy as shown by exposure of male rats in utero to the aromatase inhibitor 1-4-6 androstatriene 3-17 dione (ATD) via prenatal maternal injections from day 10 to 22 of pregnancy (Gladue & Clemens, 1980). Males, prenatally exposed to ATD, were castrated in adulthood and tested for the display of masculine behaviour in response to replacement therapy with testosterone propionate. They showed impairment of copulatory behaviour and diminished probability of ejaculating compared with control animals. Male hormone does not appear to be necessary for organizing male sexual behaviour after birth. Male rats castrated the day of birth display with subsequent testosterone replacement therapy higher mounting rates than males castrated later. However, males castrated immediately after birth displayed virtually no intromission or ejaculation (Grady & al., 1965). The important question is thus whether incomplete penis development can entirely account for the failure of neonatally castrated animals to achieve intromission or ejaculation or whether the effect is also due to incomplete neural differentiation. Numerous data indicate (Whalen, 1968) that manipulations which alter the sensitivity of the penis reduce the probability of intromission and ejaculation. On the other hand, neonatally castrated male rats may be considered to be sexually arousable, since they do respond to females and do display mounting responses. In the rhesus monkey we have indirect evidence that masculinization also occurs before birth (Phoenix & al., 1983, 1984 ; Pomerantz & al., 1985). Gender identity and sexual behaviour appeared to be masculine in females exposed to androgen during foetal life.

Defeminization

Even though androgens neonatally secreted by the testes may not be involved in masculinizing processes, there is strong evidence suggesting that they exert defeminizing effects in the male rat. This has been demonstrated in male rats castrated immediately or early after birth. When treated with oestradiol/progesterone in adulthood these animals displayed a high level of lordosis in response to male mounts (Grady & al., 1965 ; Feder & Whalen, 1965 ; Fadem & Barfield, 1981). Similarly, male rats given ATD (Fadem & Barfield, 1981) during the early neonatal period, and thus deprived of androgen, show proceptive behaviour and lordosis comparable to those of normal females after oestrogen and progesterone treatment in adulthood. However defeminization in the rat may occur earlier, i.e. prior to birth. When exposed to flutamide from the 10th to the 22nd day of pregnancy (Gladue & Clemens, 1982), adult male rats showed higher levels of feminine behavioural responses to oestrogen and progesterone than did controls. ATD treatment before birth (Whalen & Olson, 1981) also caused some degree of feminization in Sprague Dawley male rats. But there are some strain or species differences in the timing of defeminizing processes. Whalen et al. (1986) reported that male rats of the Long-Evans strain were more responsive to the effects of ATD before birth than Sprague Dawley male rats, but, in the latter, castration within 24 hours after birth strongly enhanced feminization caused by prenatal exposure to the aromatase inhibitor. Similarly, in the ferret, prenatal neural aromatic conversion of androgen to oestrogen normally initiates the process of proceptive defeminization during the last 11 days of gestation (Baum & Tobet, 1986) but its seems likely that this process is subsequently completed by the action of testicular hormones over an extended neonatal period (Baum & al., 1985). In pigs (Ford & Christenson, 1987) proceptive defeminization of the male depends on androgen acting over an extended neonatal period. In male monkeys indirect evidence supports the assumption that defeminization probably takes place before birth. Females exposed to either testosterone or dihydrotestosterone displayed receptive and proceptive behaviour at lower levels than control females and at higher levels than males castrated neonatally (Pomerantz & al., 1985). However, according to Thornton and Goy (1986) defeminization is completed during the first month after birth. Surprisingly the results of perinatal testosterone manipulation in the male vole (Microtus ochrogaster) did not support the organizational hypothesis of sexual differentiation (Petersen, 1986), since adult male voles did not show feminine responses to oestrogen even when deprived of androgen exposure by castration within three hours after birth or by perinatal treatment with either flutamide or ATD. It is possible that the prenatal critical period for defeminization in the male is more precocious. It is worthwhile mentioning the observations of Smale & al. (1985) who claimed that androgenization shortly after birth caused defeminization in the female vole while castration of males in adulthood did not result in feminine responses to oestrogen.

Independence of the Organization of Masculine and Feminine Behaviours

Although defeminization and masculinization may proceed concomitantly at least before birth in male rats and male monkeys, the potential to display masculine and feminine behaviour probably develops independently because the period of defeminization is prolonged after birth especially in the rat. Male rats born from mothers given cyproterone on day 18 of pregnancy display impaired, masculine behaviour but no signs of feminization (Perakis & Stylianopoulou, 1986). Conversely, treatment of newborn male rats with ATD from day 2 to day 10 after birth induced feminization without affecting masculinization in adulthood (Davis & al., 1979). Hormonal manipulation of foetal or neonatal females further supports the independence of organization of male and female sexual behaviour. Female rhesus monkeys rendered pseudohermaphroditic by maternal androgen administration during pregnancy maintained feminine proceptive and receptive behavioural patterns despite masculinization of play and sexual behaviours (Phoenix & al., 1983, 1984). Similarly, masculinization resulting from testosterone implantation into the preoptic anterior hypothalamic area of neonatal female hamsters appeared to be compatible with the expression of feminine behaviour in adulthood (Hamilton & al., 1981).

Impermanence of the Organizational Effects of Androgens on Brain Structures Controlling Male Sexual Behaviour

Over the two past decades several experimental results with non- manipulated animals have been reported which are in conflict with the hypothesis that the defeminizing effects of androgens during a critical prenatal and/or postnatal period are permanent in adult animals.

Experimental Display of Heterotypical Behaviour in Male Rats Castrated as Adults

For twenty years the possibility of inducing lordosis has been shown to be present in males castrated after puberty and given either successive doses (Davidson, 1969 ; Davidson & Levine, 1969 ; Arèn-Engelbrektsson & al., 1970 ; Schaeffer & al., 1986a) or a single dose (Van de Poll & Van Dis, 1977 ; Chabli & al., 1985) of oestradiol benzoate. As in females, progesterone of exogenous or adrenocortical endogenous origin is capable of potentiating the effects of oestrogen (Van de Poll & Van Dis, 1977 ; Schaeffer & Aron, 1981 ; Chabli & al. 1985 ; Schaeffer & al., 1986a). This is in contrast with previous reports which indicated that exogenous progesterone failed to facilitate the display of heterotypical behaviour in castrated rats primed with repeated doses of oestrogen (Davidson, 1969 ; Arèn-Engelbrektsson & al., 1970). Perhaps the facilitating action of progesterone could not be demonstrated in these experiments because of the high dose of oestrogen and the length of treatment ? More recently, Hennessey & al. (1986) and McEwen (1988) claimed that, in contrast with the situation in oestrogen-primed females, males are especially insensitive to the effects of progesterone in that the latter hormone was incapable of increasing the lordosis response following oestrogen-priming. In view of our own results and those of Van de Poll and Van Dis (op. cit.), we believe that some imbalance between the doses of oestrogen and progesterone used by these investigators may account for the sex difference in lordosis induction by progesterone. In any case more recent evidence strongly suggests that, despite the process of behavioural organization, the neuroendocrine substrate for feminine sexual behaviour remains sensitive to hormonal activation in adulthood.

Spontaneous Display of Heterotypical Behaviour in Male Rats and Monkeys

Until the sixties little attention was paid to bisexual behaviour in intact male rats. Stone (1924) described two males and Beach (1938) an additional two males which displayed both mounting behaviour and lordosis. In fact the intact male rat has been shown to be capable of showing lordosis in response to either flank-perineum manual stimulation or to male mounting (Södersten & al., 1974). However, there are obvious strain differences in the display of lordosis by intact male rats. In a Dutch strain of rats, Södersten & al. (1974) were unable to reproduce the effects of manual stimulation or mounting which had been previously observed in a Danish strain by Södersten and Larsson (1974). The Wistar rats bred in our colony may also display lordotic responses when mounted by a sexually active male congener but this heterotypical behaviour is rather infrequent (unpubl. observ.). Observations in subhuman primates indicate that presentation (the receptive posture of the female) may be displayed by the intact adult male macaque (Phoenix & al., 1967). Very recently spontaneous lordosis has been reported in the guinea pig (Thornton & al., 1987).

IS THE SPONTANEOUS EXPRESSION OF HETEROTYPICAL BEHAVIOUR IN THE MALE RELATED TO DEFECTS IN ORGANIIZATIONAL OR ACTIVATIONAL PROCESSES ?

There is evidence to support the notion that impairment of testosterone secretion during foetal life accounts for the display of heterotypical sexual behaviour in the rat. Heterotypical behavioural patterns were observed in adult male rats which had been exposed to prenatal stress (Ward, 1972 ; Götz & Dörner, 1980), and androgen deficiency during perinatal life has been regarded as the causal factor in the development of heterotypical sexual behaviour in these animals (Ward & Weisz, 1980 ; Dörner & al., 1983a). However, in the absence of conclusive data on the biochemical aspects of hypothalamic androgen deficiency before birth (Weisz & al., 1982) further information is required to validate this hypothesis.

New insights are also necessary, since the situation concerning a possible incidence of hormonal imbalance on the display of heterotypical behaviour in the male is far from clarified. Södersten & al. (1974) in the rat and Thornton & al. (1987) in the guinea pig could not demonstrate differences in plasma testosterone and oestradiol concentrations in males which responded by lordosis to male mounts and in those which did not. Neither did plasma testosterone concentration decrease in male homosexual subjects (Vague & Favier, 1977). These observations are not surprising in the light of the effects of steroid hormones on sexual behaviour. It is well known that testosterone induces heterotypical sexual behaviour in males castrated at birth while it provokes male sexual behaviour in animals castrated after puberty (Dörner & Hinz, 1968). Similarly oestradiol may exert either feminizing (Davidson, 1969) or masculinizing (Södersten, 1973) effects in castrated male rats depending on the sex of their mating partners. In subhuman primates the display of heterotypical behaviour in the male is independent of any gonadal hormone. This has been shown in the macaque which is capable of displaying the receptive posture of the female even after castration (Phoenix & al., 1967).

Current evidence obtained from studies in the rat supports the concept of a dual function of female hormone which might exert a heterologous function in the male in activating the neural mechanisms involved in the display of heterotypical behaviour. It is thus necessary to establish what nervous system components are involved in the expression of heterotypical behaviour in the male.

NEUROENDOCRINE MECHANISMS OF HETEROTYPICAL BEHAVIOUR IN THE MALE RAT

Assessment of Feminine Behaviour in the Male Rat

Feminine behaviour, i.e. lordosis of castrated male rats, was tested for 10 min in a mating arena with transparent plexiglass walls by placing experimental animals with two intact, adult, sexually-vigorous males. Previous observations (Schaeffer & Aron, 1981) showed that, within such a time period, the castrated males were mounted an average of 10 times by the stimulus males. Prior to testing, the stimulus males were tested for 10 min with a highly-receptive, ovariectomized female primed with oestrogen and progesterone. Only males which were known to be fully sexually active were used. They were permitted a period of 5 min to adapt to the mating arena. A castrated male was then introduced. Two indicators were used for the estimation of lordosis :

1) the proportion of animals that showed willingness to mate and thus to display at least one deep lordosis in response to male mounts. The lordosis posture consisted in arching the back, extending the neck and deviating the tail to one side thus exposing the genital region.

2) the lordosis quotient (LQ) which served as a measure of the mating performance. The LQ was computed for each castrated male rat by dividing the number of lordosis responses by the number of mounts and multiplying by 100. A mean LQ was calculated for each group of animals. Behavioural testing was performed at the end of the light period of an artificial lighting rhythm (lights on 02.00 - 16.00 hr).

Is the Hypothalamic Ventromedial Nucleus a Primary Site for the Activation of Feminine Behaviour in the Male ?

It is well-known that the ventromedial hypothalamic nucleus (VMN) plays a key role in the control of lordosis behaviour in the female rat (Carrer & al., 1973 ; Kennedy & Mitra, 1963 ; Pfaff & Sakuma, 1979). Recently, Okada & al. (1980) and Yamanouchi and Arai (1983) reported the existence of a dual system facilitating the display of female sexual behaviour, involving both the VMN and the pontine periventricular gray. It was thus important to determine in the male whether the VMN represented a primary site for the activation of heterotypical sexual behaviour. This was first suggested by Davis and Barfield (1979) who reported activation of feminine sexual behaviour by implants of oestradiol benzoate in the VMN of castrated males primed with oestrogen. Recently the effects of VMN lesions on the display of heterotypical sexual behaviour have been studied (Chateau & al., 1987) using male rats castrated as adults and subsequently given oestradiol benzoate and progesterone. Table 1 shows that VMN lesions suppressed the display of heterotypical sexual behaviour compared with sham and unoperated animals.

The VMN may thus be considered as a target area for gonadal hormones in the stimulation of lordosis responses in the male, as is the case in the female rat. However a possible influence of other central nervous structures in the regulation of heterotypical sexual behaviour of the male rat cannot be disregarded. Very recently, Yamanouchi and Arai (1985) showed in male rats feminized with ovarian hormones that after roof anterior deafferentation, the pontine periventricular gray played a critical role in the facilitation of the lordosis reflex and constituted the target for an extrahypothalamic system independent of the VMN facilitory mechanisms. Further studies are therefore necessary to establish the respective involvement of both the VMN and the pontine periventricular gray in the control of heterotypical sexual behaviour in the male rat.

Table 1   Effects of VMN lesions on lordosis behaviour in rats castrated as adults and given 25 μg oestradiol benzoate and 1 mg progesterone (P) at a 39 hr interval.

Treatment *	Proportion of rats displaying lordosis	Mean lordosis quotient (LQ) in rats displaying lordosis
VMN lesions	0/16	-
Sham VMN	8/16	36.9 ± 3.1
DMN** lesions	8/16	45.8 ± 11.9

* Testing with two vigorous males was performed within 8 hr after P injection.
** Hypothalamic dorsomedial nucleus.

VMN lesions vs sham and DMN lesions (P < 0.001).

Involvement of the Amygdala in the Control of Feminine Behaviour in the Male

Until recently few studies have concerned the function of the amygdala in the control of sexual behaviour in the female rat. More than 20 years ago De Groot and Critchlow (1960) and Zouhar and De Groot (1963) reported that lesions of the mediobasal amygdala increased mating frequency in cyclic pregnant female rats. These findings disagreed with those of Eleftheriou and Zolovick (1966) who tested the effects of basolateral and ventromedial amygdaloid lesions in the cyclic deermouse. They found that the former rendered the females more receptive and that the latter provoked opposite effects. An elegant study by

Masco and Carrer (1980) showing the effects of either lesioning or electrochemical stimulation of different amygdaloid nuclei has clarified the situation. Lesions placed in the anterior part of the corticomedial amygdaloid nucleus (CMN) were shown to decrease LQ values in ovariectomized rats given oestradiol benzoate and progesterone, while electrochemical stimulation had the opposite effect. The CMN thus appears to exert a facilitatory influence on sexual behaviour in the female rat.

Such control of lordosis by the CMN was even more plausible since the VMN, which constitutes in the female a target for the activation of lordosis by gonadal hormones (Clark & al., 1981 ; La Vaque & Rodgers, 1975 ; Mathews & Edwards, 1977), receives projections from the CMN (see for review De Olmos & al., 1985). These data, encouraged us to investigate whether the amygdala is involved in the mechanisms controlling heterotypical sexual behaviour in the male, since the VMN is now known (Chateau & al., op. cit.) to be absolutely necessary for the display of lordosis in the male rat.

Recent findings (Chateau & Aron, 1988) support this hypothesis. They demonstrate that the amygdala exerts modulatory effects on lordosis in the male as well as the female. Extended stereotaxic lesions of the CMN completely suppressed lordosis in rats castrated as adults and subsequently given 25μg oestradiol benzoate and 150μg progesterone. There is thus good evidence that a common facilitatory amygdaloid ventromedial hypothalamic pathway exists in both the male and female for the control of lordosis.

However, the CMN is not the only amygdaloid structure involved in the control of lordosis in the female. The amygdaloid complex may be regarded as a functionally heterogeneous structure with different nuclei exerting opposing effects on the same function. Masco and Carrer (1980) reported that lesions to the posterior part of the lateral amygdaloid nucleus (LN) significantly increased sexual receptivity of ovariectomized rats given ovarian hormones. Similarly, we induced "hypersexuality" (demonstrated by an increase in the number of animals displaying willingness to mate and by enhancement of mating performance) in LN-lesioned castrated male rats primed with oestradiol and progesterone.

The posterior part of the LN may thus be considered in the male (as in the female), as an inhibitory structure for the expression of feminine behaviour. It must be emphazised that a syndrome of hypersexuality such as previously described by Klüver and Bucy (1939) in the monkey and by Schreiner and Kling (1953) in the cat was not observed before earlier males feminized with ovarian hormones. However, it should be remembered that LN lesions only exert delayed inhibitory effects on lordosis after surgery. Probably a recuperative reorganizational process (Clemens, 1978) of other amygdaloid structures is necessary to allow the display of "hypersexuality" in LN lesioned animals.

Impairment of lordosis in castrated rats by lesions in the anterior part of the LN support this interpretation. Thus the LN constitutes a heterogeneous structure with a posterior and an anterior region exerting opposite effects on the display of lordosis. This explains why extensive lesions of the LN completely failed to affect lordosis in castrated rats feminized with ovarian hormones (Chateau & Aron, 1988). The question now is how LN neurons interact in the mechanisms which control lordosis in the male. The anterior and posterior part of the LN are known to be poorly interconnected and it is questionable whether the posterior part of the LN exerts its inhibitory effects on the anterior part of the nucleus. The pathway involved in the control of lordosis by the anterior part of the LN must also be determined. Anatomical (Kita & Oomura, 1982 ; Zaborszky, 1982 ; Luiten & al., 1983) and electrophysiological (Dreifuss, 1972 ; Murphy, 1972 ; Renaud & Martin, 1975) data suggest that the VMN receives projections from the LN. On the other hand, Ottersen (1982) described projections from the posterior part of the LN to the CMN. Therefore extra-amygdaloid and intra-amygdaloid connections exist which may account for the control exerted by the LN on lordosis in the male rat. In any case it may be suggested that a common behavioural control system involving both amygdaloid nuclei and the hypothalamic VMN underlie the display of lordosis in the male and female rat.

OLFACTORY CONTROL OF FEMININE BEHAVIOUR IN THE MALE RAT

General Background

Several years ago, Beach (1976) proposed a concept for the sexual behavioural characteristics of female mammals. He discriminated between "attractivity" which is the male appetitive reaction of approaching and investigating the female, and "proceptivity" which represents an appetitive response of the female to male stimuli. Receptivity, which constitutes the final consummatory phase of mating may be defined as the final readiness to allow copulation.

Undoubtedly chemical communication is important in the establishment of these sexual interactions (for review see, Aron, 1979). In rodents olfactory signals play a major role when females are attractive to a male or attracted by them. Other sensory signals may, however, be involved as far as higher mammals are concerned. Visual cues increase sexual attractivity in chimpanzees. Auditory, visual and even tactile stimuli originating from the boar facilitate the standing reaction (the receptive posture) in the gilt. Gustatory stimuli received by the males, when they lick the anogenital region, contribute to the sexual attractiveness of the female in various species of mammals. Consummatory behaviour does not escape the control of sensory stimuli. This was observed in the rat when genital manipulations evoked the postural posture of lordosis. Genital stimulation associated with intromission has been considered to facilitate lordosis an effect prevented by genital denervation. However the role played by olfactory stimuli in the execution of consummatory behaviour should not be ignored. Observations in our laboratory have demonstrated (Schaeffer & al., 1982) that the consummatory phase of mating is under olfactory control in the female rat. These findings focus attention on the possible involvement of the olfactory system in the control of lordosis in the male. The experiments reported below, showing that olfactory signals modulate the display of lordosis in the male rat, strongly support this assumption.

Influence of Olfactory Signals on Lordosis Behaviour in the Male Rat

For some years we speculated (Schaeffer & Aron, 1980 ; 1981 ; 1982) that olfactory stimuli provided by the male influence the display of feminine behaviour in castrated congeners primed with ovarian hormone. Urine from intact male Wistar rats was selected as a source of pheromones. Male rats were castrated as adults and placed on a grid 5 cm above the floor or their cages. They were given 75 μg oestradiol benzoate and 1 mg progesterone (P) at an interval of 40 hr. Sexual behavioural testing was performed within 8 $\frac{1}{2}$ hours after P injection. One group of animals was exposed to the odour of urine by sprinkling the litter of their cages with 5 ml urine at the time of P injection. Other groups were subjected to either complete or anterior olfactory bulb removal prior to behavioural testing to determine the influence of the olfactory system on lordosis. Animals which were neither operated on nor exposed to the odour of urine served as controls.

Table 2 shows that the number of animals displaying lordosis was higher in rats exposed to male urine odour than in controls. Complete olfactory bulb removal mimicked the effects of the olfactory signals. In contrast, anterior olfactory bulb removal did not affect lordosis behaviour compared with controls. Statistical analysis confirmed these data ($P > 0.05$).

No difference was noted in LQ values between the complete and anterior bulbectomized animals. In combination those groups displayed high LQ values compared with animals exposed to male urine odour and controls ($P < 0.01$). These results indicate that the olfactory bulbs, on their own, can inhibit the expression of feminine behaviour in the male rat. This inhibitory control seems to depend on the caudal region of the olfactory bulbs, since anterior olfactory bulb removal leaving posterior olfactory bulb undamaged did not facilitate the display of lordosis compared with complete olfactory bulb removal. This led us to suggest that the accessory olfactory bulbs (AOB) which are located in the caudal part of the olfactory bulbs, constitute likely candidates to exert

Table 2   Effects of the olfactory system on the display of
          lordosis in castrated male rats given ovarian hormones

Treatment	Proportion of rats displaying lordosis	Mean LQ in rats displaying lordosis
Controls	22/60	48 $\pm$ 6
Complete olfactory bulb removal	36/58	65 $\pm$ 5
Anterior olfactory bulb removal	18/75	58 $\pm$ 7
Exposure to male urine odour	19/30	47 $\pm$ 6

such inhibitory effects. This hypothesis was verified in a series of
experiments (Schaeffer & al., 1986b) where the display of lordosis
was studied after complete AOB removal. The schedule of
hormonal treatment and of sexual behavioural testing was
identical to that in the preceding experiments. Sham operated and
unoperated animals served as controls. After AOB removal a significant rise in the proportion of animals (15/17) displaying lordosis
was observed compared with sham (16/27) and unoperated (19/35)
animals, but LQ values did not differ in the three groups of
animals. It would thus be reasonable to suggest that the accessory
olfactory system inhibits the expression of feminine behaviour in
the male rat and that olfactory signals from the male release this
inhibition. However it is important to note that the modulatory
influence of the AOB on lordosis was only apparent when the
effects of male urine on mating frequency were considered. In
contrast neither olfactory stimuli originating from the male, nor
accessory olfactory bulb removal affect the mating performance.
This was rather unexpected since complete bulb removal and
anterior olfactory bulb removal have been shown to enhance the
mating performance. It is therefore tempting to suggest that the
principal olfactory pathway conveys olfactory cues which modulate
the sexual performance. The mechanisms by which males are
attracted to female vaginal secretions in the hamster are relevant
to this problem. Both the accessory and the main olfactory
systems are known to play an important role in sexual interactions
between male and female hamsters (Powers & al., 1979). The male
attraction to female vaginal secretions depends on both the sense
of smell and contact perception by the vomeronasal organ which

projects fibres to the AOB. Further experiments are necessary to determine more accurately the involvement of the main olfactory system in the modulation of the mating performance in feminized male rats. We are nevertheless able, on the basis of the most recent data on the neuroendocrine mechanism governing lordosis in the male rat, to formulate a tentative interpretation of the modulation of lordosis behaviour by the olfactory system in the male. According to Yamanouchi and Arai (1985) the pontine periventricular gray plays a critical role in the facilitation of the lordosis reflex in the male and constitutes the target for an extrahypothalamic inhibitory system independent of the VMN, and probably located in the septum and/or the preoptic area. We may question whether the extrahypothalamic inhibitory system postulated by these authors does not correspond to the AOB which would exert its influence via the corticomedial amygdaloid nucleus on the hypothalamic VMN and perhaps also on the pontine periventricular gray. In fact, very recent findings in our laboratory rather suggest a dual inhibitory control involving both the AOB and another limbic structure on the display of lordosis by the male rat. Obviously, further neuroendocrinological approaches are required to investigate this problem.

Emission and Perception of the Olfactory Signals which Modulate Lordosis in the Male

Information on the chemical nature of the pheromones which act as sex attractants in the rodents is scarce. Dimethyldisulphide, a substance known to be attractive to the male, has been isolated from vaginal secretions in the female hamster (Singer, 1976) and aphrodisine, a water-soluble protein in hamster vaginal discharge has been purified and shown to stimulate copulatory behaviour in the male hamster (Singer & al., 1986). In male mice, the preputial glands constitute a source of the pheromones (Bronson & Caroom, 1971) which are attractive to the females. Factors produced by the preputial gland are present in urine and their activity has been attributed to a free fatty acid fraction (Gaunt, 1968). In the rat, the male attractiveness to the female, which is suppressed by castration (Carr & Caul, 1962) depends on the pheromonal activity of urine. The discovery that the odour of male urine could facilitate the expression of lordosis in the male rat encouraged us to suggest that the accessory reproductive glands produced pheromones responsible for this effect. This was strengthened by data showing that the facilitatory effects of male urine are androgen-dependent (Schaeffer & Aron, 1982). It is well known that the function of the accessory reproductive glands in the male depend on testicular endocrine activity (Price & Williams-Ashman, 1961). To test our hypothesis, we collected urine from intact male rats and from castrated male rats supplemented with testosterone

(Chabli & Aron, 1986). The animals were previously deprived of the preputial and/or the coagulatory glands. Pheromonal activity disappeared when urine was collected from animals deprived of both their preputial and coagulatory glands. In contrast, urine retained its pheromonal activity after removal of one of these glands. These glands may therefore substitute for each other and release urinary pheromones capable of facilitating lordosis in feminized male rats. The substances responsible for this pheromonal activity remain to be identified.

The question also arises as to whether the perception of the male olfactory signals by the feminized animals is hormone dependent. It is worth noting that both oestradiol benzoate and progesterone (P) have been used to feminize males (Schaeffer & Aron, 1981, 1982). It was therefore necessary to determine whether EB and/or P were necessary for the olfactory signals to exert their facilitatory effects on feminine behaviour. Different experimental models were used to investigate this problem (Chabli & al., 1985). Some castrated males were given either a single dose or successive daily doses of EB and exposed to the odour of male urine within 6 - 1 hr before behavioural testing. No pheromonal effect was observed under these experimental conditions. Using 75 μg EB, 39 hr prior to a dose of 1 mg P (capable on its own of potentiating the effects of EB on lordosis) we exposed the animals to male urine odour at different times during pheromonal treatment. Exposure to urine during oestrogen treatment remained ineffective but significantly increased the number of animals showing lordosis when performed at the time of the P injection. It was also necessary to evaluate the requirement of P for the facilitation of lordosis by urine odour. Interestingly we observed that a dose of P unable to facilitate the effects of EB on lordosis, rendered the animals responsive to the odour of urine (Table 3).

These observations led us to suggest that both the emission and the perception of the signals provided by the male are under hormonal control. However P appeared to play a key role in the facilitation of lordosis by olfactory signals since the odour of male urine was only efficient in the animals subsequently given EB and P. The presence of oestrogen and progesterone receptors in the mediobasal hypothalamus in both sexes is well documented (Muldoon, 1980). Studies on the female rat indicate that oestradiol may cause an increase in the concentration of hypothalamic cytoplasmic P receptors (McLusky & McEwen, 1978). A good correlation has also been established between the induction of lordosis by oestradiol in ovariectomized rats and the rise in cytoplasmic P receptors in the mediobasal hypothalamus (Parsons & al., 1980).

If we consider that the VMN constitutes in the male (Davis & Barfield, 1979 ; Chateau & al., 1987), as in the female (Dörner & al., 1968), a target for the induction of lordosis by ovarian

Table 3   Effects of exposure to male urine odour on the display
           of lordosis in male rats castrated as adults and given
           oestradiol benzoate (EB) and progesterone (P) 39 hr
           apart.

	Treatment[+]	Proportion of rats displaying lordosis
25 μg EB	no urine urine [*]	6/30 5/30
25 μg EB + 100 μg P	no urine urine[*]	4/28 20/30 ◊
25 μg EB + 150 μg P	no urine urine[*]	12/29 13/16 §

+ Testing with two vigorous males was performed within 48 ± 1 hr
  EB injection.
* Bedding soiled with 5 ml urine within 39 hr after EB injection.
◊ P < 0.001 vs no urine value.
§ P < 0.01 vs no urine value.

hormones, it is reasonable to assume that the olfactory signals
interplay with the hormonal signals at the hypothalamic level.
Recent data (Samama & Aron, 1989) showing in male rats primed
with EB and P a close relationship between the facilitatory
effects of olfactory signals on lordosis and the concentration and
binding capacity of E_2 nuclear receptors in the mediobasal
hypothalamus support this assumption. These results are important
because they provide the first evidence that certain molecular
mechanisms are involved in the effects of sensory stimuli on the
nervous structures which subserve the display of sexual behaviour.

Lordosis in the Male Rat : Phantasy or Reality

We must obviously bear in mind that castrated rats primed
with ovarian hormones represent an artificial model to study the
influence of the olfactory system on the display of lordosis in the
male. On the other hand, only a small number of sexually inex-
perienced intact males in our colony respond to male mounts by
lordosis in the absence of any hormonal treatment. Why do very
few animals display lordosis spontaneously ? Are they submitted to
the repressive action of the AOB ? Alternatively, are the
olfactory signals from the male unable to release this inhibition

due to hormonal conditions which are not compatible with the perception of these signals ? Experiments are currently in progress which afford a preliminary response to these questions. In accordance with previous observations made on the rat (Södersten & al., 1974) and the guinea pig (Thornton & al., 1987), oestradiol and testosterone values did not differ among male rats which, in our colony, displayed or failed to display spontaneous lordosis behaviour. In the latter the hormonal milieu was likely to be incompatible with activation of the nervous structures subserving lordosis, since activational processes are required for the expression of sexual behaviour. The brain lack of sensitivity to oestradiol also probably explains why accessory olfactory bulb removal had no facilitatory effects on the release of lordosis behaviour in intact animals (unpubl. observ.) which did not display spontaneous lordosis in response to male mounts. Recently, however, we observed that blood progesterone levels were higher in sexually experienced male rats displaying lordosis in response to male mounts than in those which did not. Thus the possibility cannot be ruled out that progesterone-induced sensitization to male's olfactory cues allowed the expression of lordosis behaviour in intact animals as in castrated animals primed with oestrogen (Chabli & al., 1985). Further investigations are necessary for a more complete understanding of the neuroendocrine mechanisms which allow some male rats to spontaneously display feminine behaviour. Sufficient data are nevertheless available at the present time to suggest that some interactions between organizational, activational and pheromonal processes are involved in the spontaneous expression of lordosis, in the male rat.

PHYSIOPATHOLOGICAL DATA

Sexual Orientation in Male and Female Pseudohermaphrodites

Several years ago Imperato-McGinley & al. (1974, 1979) reported their observations in male pseudohermaphrodites born in the Dominican Republic. All suffer from α-reductase deficiency. The babies are born with genitalia more female than male, because, in the absence of conversion of testosterone to dihydrotestosterone, the accessory reproductive glands and the genitalia fail to develop. These individuals are raised as females until adolescence and they become sufficiently virile by the age of 16 years to identify themselves as males at that time. Thus these subjects have changed both gender identity and gender role compared to those of sexually active males attracted to females. In support of the findings of Imperato-McGinley and her colleagues, Gajdusek (1977) reported another form of male hermaphroditism occurring in New Guinea. In accordance with the different customs of this country, the affected children are either accepted and reared as males in their society or identified at birth as females. In the first case the tribe creates a myth to explain the phenomenon. In the second, the subjects are reassigned as males when they become virile during adolescence. If we postulate that the human male is subject to the organizational effects of androgens before birth, masculinization of the central nervous system, which depends on aromatization of androgens, certainly has taken place in male pseudohermaphrodites. This would mean that, under certain favourable cultural circumstances the prenatal organizational effects of androgens may overcome the effects of the environmental factors which should have orientated gender identity after birth in a direction opposite to the genetic sex.

However, environmental influences should not be under-estimated as demonstrated in a pathological condition, namely congenital adrenal hyperplasy (CAH), that results in elevated androgens during prenatal life in genetic females. This condition is related to a defect in 21-hydroxylase affecting the functioning of the adrenal cortex so that it cannot produce cortisol. Females with this abnormality are born as female pseudohermaphrodites with masculinized external genitalia. Postnatal treatment consists of corticosteroid replacement therapy and correction surgery to restore the external genitalia to a normal appearance. Generally speaking, none of the data concerning CAH support hypotheses suggesting a determining role of prenatal androgens on later sexual orientation (Baker, 1980). This role is unlikely, in view of the number of foetally masculinized girls who have been reported to be married or involved in normal heterosexual behaviour. Since ACH subjects have been submitted to an excess of androgens before birth we must conclude that organization of male behaviour may be impermanent in the human as it is the case in the rat and in same other species. The present state of our knowledge does not allow us to disregard either the environmental or or-ganizational factors in the determination of sexual orientation in the human.

Stress and Sexual Orientation

Data collected (Dörner & al., 1980) in men who were born in Germany before, during and after the second world war suggest that stressful events in prenatal or early postnatal life may represent a risk factor for homosexuality in human males. This was recently confirmed (Dörner & al., 1983b) in bi-or-homosexual men who were asked about maternal stressful events that may have occurred during their prenatal life. An increase in the incidence of prenatal stress was found particularly in homosexual men. Thus we cannot rule out the possibility that a defect in organization of male sexual behaviour before birth may be responsible for the display of heterotypical sexual behaviour in human beings.

CONCLUSIONS

The similar influence of stressful events occurring during prenatal life on the sexual orientation of male rats and humans suggest that human sexual behaviour is fundamentally determined by biological mechanisms involving the organizational effects of androgens on the central nervous system. However certain physio-pathological data emphasize the importance of environmental factors in the determination of sexual orientation in humans. Obviously, educational and social factors are significant para-meters in CAH girls who are raised as females after prenatal exposure to an excess of androgens (Money & Ehrhardt, 1972). In this case, the effects of possible masculinization before birth is overcome by postnatal social and educational environmental influences. Conversely, the psychological development of subjects showing 5α-reductase deficiency before birth clearly indicates that the determination of gender identity during infancy, defined as persistent experience of oneself as male or female, is not permanent. These data raise the question about the hypothesis (Baker, 1980) concerning the determination of gender identity before the age of 2 years. The influence of environment on sexual orientation is not only characteristic of human beings. The fact that male olfactory signals can modulate the display of feminine behaviour in male rats which have been submitted before birth to the organizational effects of androgens suggests that sexual orientation in rodents may depend on the balance between organizational and environmental factors.

It is of course not possible to use rodents as an animal model for psychological studies of sexuality as is the case for human beings. Nevertheless the use of such a model cannot be disregarded since it leads to a better understanding of human sexuality. Briefly, it may be said that sexuality in rodents is likely to be under more rigid hormonal control than is the case in the human. The influence of sensory environment on the determination of sexual orientation is thus rather dependent upon the hormonal state of the animal. In contrast, in the human, the development of "erotosexuality", to use the terminology proposed by Money (1981) is mainly subordinated to the stimulus input on the brain of the educational and social environment. Thus parental influence may be of paramount importance in the determination of sexual orien-tation in both sexes. Explanations based on simple cause-effect, however, do not account for the extreme complexity of the mechanisms of human sexuality. Cultural influences appear to be determinant for sexual orientation in adolescence and we cannot ignore this factor for the biological interpretation of the mechanisms of sexuality. As long as investigative approaches that are ethically acceptable for human experimentation remain undiscovered, rodents and other mammalian species will continue to constitute the only experimental basis permitting tentative interpretations of neuroendocrine mechanisms which might con-tribute, before birth, to the organization of sexual behaviour in human beings. Environmental factors certainly play a major role in the sexual orientation of human individuals but we have no reason, in the absence of any other evidence to disregard the biological background of behavioural bisexuality in the human.

REFERENCES

Arèn-Engelbrektsson, B., Larsson, K., Sodersten, P. and Wilhelmsson, M. (1970). The female lordosis pattern induced in male rat by estrogen. Hormones and Behavior, 1, 181-188.

Aron, Cl. (1979). Mechanisms of control of the reproductive function by olfactory stimuli in female mammals. Physiological Reviews, 59, 229-284.

Baker, S.W. (1980). Psychosexual differentiation in the human. Biology of Reproduction, 22, 61-72.

Baum, M.J. and Tobet, S.A. (1986). Effect of prenatal exposure to aromatase inhibitor testosterone or antiandrogen on the development of feminine sexual behavior in ferrets of both sexes. Physiology and Behavior, 37, 111-118.

Baum, M.J., Stockmann, E.R. and Lundell, L.A. (1985). Evidence of proceptive without receptive defeminization in male ferrets. Behavioral Neurosciences, 99, 742-750.

Beach, F.A. (1938). Sex reversals in the mating pattern of the rat. Journal of Genetical Psychology, 53, 329-334.

Beach, F.A. (1976). Sexual attractivity, proceptivity and receptivity in female mammals. Hormones and Behavior, 7, 105-138.

Bronson, F.H. and Caroom, D. (1971). Preputial glands of the male mouse : attractant function. Journal of Reproduction and Fertility, 25, 279-282.

Carr, W.J. and Caul, W.F. (1962). The effect of castration in the rat upon discrimination of sex odors. Animal Behavior, 10, 20-27.

Carrer, H., Asch, G. and Aron, C. (1973). New facts concerning the role played by the ventromedial nucleus in the control of estrous cycle duration and sexual receptivity in the rat. Neuroendocrinology, 13, 129-158.

Chabli, A. and Aron, C. (1986). Mechanisms of emission of the olfactory signals which facilitate feminine behavior in the rat. Biology of Behaviour, 11, 61-69.

Chabli, A., Schaeffer, C., Samama, B. and Aron, C. (1985). Hormonal control of the perception of the olfactory signals which facillitate lordosis behavior in the male rat. Physiology and Behavior, 35, 729-734.

Chateau, D. and Aron, C. (1988). Heterotypic sexual behavior in male rats after lesions in different amygdaloid nuclei. Hormones and Behavior, 22, 379-388.

Chateau, D., Chabli, A. and Aron, C. (1987). Effects of ventromedial nucleus lesions on the display of lordosis behavior in the male rat. Interactions with facilitory effects of male urine. Physiology and Behavior, 39, 341-345.

Clark, A.S., Pfeifle, J.K. and Edwards, D.A. (1981). Ventromedial hypothalamic damage and sexual proceptivity in female rats. Physiology and Behavior, 27, 597-602.

Clemens, L.G. (1978). Neural plasticity and feminine behavior in the rat. In McGill T.E., Dewsbury D.A. and Sachs B.D. (Eds), Sex and Behavior, pp. 243-266, Plenum Press, New York.

Corbier, P., Kerdelhue, B., Picon, R. and Roffi, J. (1978). Changes in testicular weight and serum gonadotrophin and testosterone levels before, during and after birth in the perinatal rat. Endocrinology, 103, 1985-1991.

Csernus, V. (1986). Production of sexual steroids in rats during pre- and -early postnatal life. Experimental and Clinical Endocrinology, 88, 1-5.

Davidson, J.M. (1969). Effects of estrogen on the sexual behavior of male rats. Endocrinology, 84, 1365-1372.

Davidson, J.M. and Levine, S. (1969). Progesterone and heterotypic sexual behaviour in male rats. Journal of Endocrinology, 44, 129-130.

Davis, P.G. and Barfield, R.J. (1979). Activation of feminine behavior in castrated male rats by intrahypothalamic implants of estradiol benzoate. Neuroendocrinology, 28, 228-233.

Davis, P.G., Chaptal, C.V. and McEwen, B.S. (1979). Independence of the differentiation of masculine and feminine behavior in rats. Hormones and Behavior, 12, 12-19.

De Groot, J. and Critchlow, V. (1960). Effect of limbic system on reproductive functions of female rats. Physiology, 3, 49 (Abstr.).

De Olmos, J., Alheid, G.F. and Beltramino, C.A. (1985). Amygdala. In Paxinos G. (Ed.). The Rat Nervous System, Vol. 1, Forebrain and Midbrain, pp. 223-334, Academic Press., New York.

Dörner, G. and Hinz, G. (1968). Induction and prevention of male homosexuality by androgen. Journal of Endocrinology, 40, 387-388.

Dörner, G., Docke, F. and Moustafa, S. (1968). Differential localization of a male and female hypothalamic mating center. Journal of Reproduction and Fertility, 17, 583-586.

Dorner, G., Geier, Th., Ahrens, L., Krell, L., Münx, G., Sieler, H., Kittner, E. and Müller, H. (1980). Prenatal stress as possible aetiogenetic factor of homosexuality in human males. Endokrinologie, 75, 365-368.

Dörner, G., Götz, F. and Döcke, W.D. (1983a). Prevention of demasculinization and feminization of the brain in prenatally stressed male rats by perinatal androgen treatment. Experimental and Clinical Endocrinology, 81, 88-90.

Dörner, G., Schenck, B., Schmiedel, B. and Ahrens, L. (1983b). Stressful events in prenatal life of bi- and -homosexual men.

Experimental and Clinical Endocrinology, 81, 83-87.

Dreifuss, J.J. (1972). Effects of electrical stimulation of amygdaloid complex on the ventromedial hypothalamus. In Eleftheriou B.E. (Ed). The Neurobiology of the Amygdala, pp. 295-317, Plenum Press, New York.

Eleftheriou, B.E. and Zolovick, A.J. (1966). Effect of amygdaloid lesions on estrous behavior in the deermouse. Journal of Reproduction and Fertility, 2, 451-453.

Fadem, B.H. and Barfield, R.J. (1981). Neonatal hormonal influences on the development of proceptive and receptive feminine sexual behavior in rats. Hormones and Behavior, 15, 282-288.

Feder, H.H. and Whalen, R.E. (1965). Feminine behavior in neonatally castrated and estrogen-treated male rats. Science, 147, 306-307.

Ford, J.J. and Christenson, R.K. (1987). Influences of pre- and postnatal testosterone treatment on defeminization of sexual receptivity in pigs. Biology of Reproduction, 36, 581-587.

Gajdusek, D.C. (1977). Urgent opportunistic observations. The study of changing, transient, disappearing phenomena of medical interest in disrupted primitive human communities. Ciba Foundation Symposium, 69-94.

Gaunt, S.L. (1968). Studies on the preputial gland as a source of a reproductive pheromone in the laboratory mouse (Mus musculus) (Ph. D. Thesis). Burlington : Univ. of Vermont.

George, F.W. and Ojeda, S.R. (1982). Changes in the aromatase activity in the rat brain during embryonic neonatal and infantile development. Endocrinology, 111, 522-529.

Gladue, B.E. and Clemens, L.G. (1980). Masculinization diminished by disruption of prenatal estrogen biosynthesis in male rats. Physiology and Behavior, 25, 589-593.

Gladue, B.E. and Clemens, L.G. (1982). Development of feminine sexual behavior in the rat : androgenic and temporal influences. Physiology and Behavior, 29, 263-267.

Gogan, F., Slama, A., Bizzini-Koutznetzova, B., Dray, F. and Kordon, Cl. (1981). Importance of perinatal testosterone in sexual differentiation in the male rat. Journal of Endocrinology, 91, 75-79.

Gorski, R.A., Gordon, J.H., Shryne, J.E. and Southam, A.M. (1978). Evidence for a morphological sex difference within the medial preoptic area of the rat brain. Brain Research, 148, 333-346.

Gorski, R.A., Harlan, R.E., Jacobson, C.D., Shryne, J.E. and Southam, A.M. (1980). Evidence for the existence of a sexually dimorphic nucleus in the preoptic area of the rat. Journal of Comparative Neurology, 193, 529-539.

Götz, F. and Dörner, G. (1980). Homosexual behaviour in prenatally stressed male rats after castration and oestrogen treatment in adulthood. Endokrinologie, 76, 115-117.

Grady, K.L., Phoenix, C.H. and Young, W.C. (1965). Role of the developing rat testes in differentiation of the neural tissues mediating matin behavior. Journal of Comparative and Physiological Psychology, 59, 176-182.

Habert, R. and Picon, R. (1984). Testosterone, dihydrotestosterone and oestradiol -17β levels in maternal and fetal plasma and in fetal testes in the rat. The Journal of Steroid Biochemistry, 21, 193-198.

Hamilton, M.A., Vomachka, A.J., Lisk, R. and Gorski, R.A. (1981). Effect on neonatal intrahypothalamic testosterone implants on cyclicity and adult sexual behavior in the female hamster. Neuroendocrinology, 32, 234-241.

Hennessey, A.C., Wallen, K. and Edwards, D.A. (1986). Preoptic lesions increase the display of lordosis by male rats. Brain Research, 370, 21-28.

Imperato-McGinley, J., Guerrers, L., Gautier, T. and Peterson, R.E. (1974). Steroid 5α-reductase deficiency in man. An inherited form of male pseudohermaphroditism. Science, 186, 1213-1215.

Imperato-McGinley, J., Peterson, R.E., Gautier, T. and Sturla, E. (1979). Androgens and the evolution of male-gender identity among male pseudohermaphrodites with 5α-reductase deficiency. New England Journal of Medicine, 300, 1233-1237.

Kennedy, G.C. and Mitra, J. (1963). Hypothalamic control of energy balance and the reproductive cycle in the rat. Journal of Physiology, London, 166, 395-407.

Kita, H. and Oomura, Y. (1982). HRP study of the afferent connections to the rat medial hypothalamic region. Brain Research Bulletin, 8, 53-62.

Kluver, H. and Bucy, P. (1939). Preliminary analysis of functions of the temporal lobe in monkeys. Archives in Neurology and Psychiatry, 42, 979-1000.

Krafft-Ebing, R.V. (1931). Psychopathia Sexualis, 16th and 17th German ed., rewritten by Moll A. (translation by Lolstein R.), 906 p., Payot, Paris.

La Vaque, T.J. and Rodgers, C.H. (1975). Recovery of mating behavior in the female rat following VMH lesions. Physiology and Behavior, 14, 59-63.

Luiten, P.G.M., Ono, T., Nishimo, H.Y. and Fukuda, M. (1983). Differential input from the amygdaloid body to the ventromedial hypothalamic nucleus in the rat. Neuroscience Letters, 35, 253-258.

MacLusky, N.J. and McEwen, B.S. (1978). Oestrogen modulates progestin receptor concentration in some brain areas, but not in others. Nature, 274, 276-278.

Masco, D.H. and Carrer, H.F. (1980). Sexual receptivity in female rats after lesion or stimulation in different amygdaloid nuclei. Physiology and Behavior, 24, 1073-1080.

Mathews, D. and Edwards, D.A. (1977). The ventromedial nucleus of the hypothalamus and the hormonal arousal of sexual behaviors in the female rat. Hormones and Behavior, 8, 40-51.

Matsumoto, A. and Arai, Y. (1986). Male-female difference in synaptic organization of the ventromedial nucleus of the hypothalamus in the rat. Neuroendocrinology, 42, 232-236.

McEwen, B.S. (1988). Genomic regulation of sexual behavior. The Journal of Steroid Biochemistry, 30, 179-183.

Money, J. (1981). The development of sexuality and eroticism in humankind. Quaterly Review of Biology, 56, 379-404.

Money, J. and Ehrhardt, A.A. (1972). Gender dimorphic behavior and fetal sex hormones. Recent Progress in Hormone Research, 28, 735-763, Academic Press, New York.

Muldoon, T.G. (1980). Role of receptors in the mechanism of steroid hormone action in the brain. In Motta M. (Ed.). The Endocrine Function of the Brain, pp. 51-93, Raven Press, New York.

Murphy, J.T. (1972). The role of the amygdala in controlling hypothalamic output. In Elephteriou B.E. (Ed.). The Neurobiology of the Amygdala, pp. 371-395, Plenum Press, New York.

Naftolin, F., Ryan, K.J. and Petro, Z. (1972). Aromatization of androstenedione by anterior hypothalamus of adult male and female rats. Endocrinology, 90, 295-298.

Naftolin, F., Ryan, K.J., Davies, I.J., Reddy, V.V., Flores, F., Petro, Z., Kuhn, M., White, R.J., Taksaka, Y. and Wolin, L. (1975). The formation of estrogens by central neuroendocrine tissues. Recent Progress in Hormone Research, 31, 295-315, Academic Press, New York.

Okada, R., Yamanouchi, K. and Arai, Y. (1980). Recovery of sexual receptivity in female rats with lesions of the ventromedial hypothalamus. Experimental Neurology, 68, 595-600.

Ottersen, O.P. (1982). Connections of the amygdala of the rat. IV : corticoamygdaloid and intraamygdaloid connections as studied with axonal transport of horse radish peroxydase. Journal of Comparative Neurology, 205, 30-48.

Parsons, B., MacLusky, N.J., Krey, L., Pfaff, D.W. and McEwen, B.S. (1980). The temporal relationship between estrogen-inducible progestin receptors in the female rat brain and the time course of estrogen activation of mating behavior. Endocrinology, 107, 777-779.

Perakis, A. and Stylianopoulou, F. (1986). Effects of a prenatal androgen peak on brain sexual differentiation. Journal of Endocrinology, 108, 281-285.

Petersen, S.L. (1986). Perinatal androgen manipulations do not affect feminine behavioral potentials in voles. Physiology and Behavior, 36, 527-532.

Pfaff, D.W. and Keiner, M. (1973). Atlas of estradiol concentrating cells in the central nervous system of the female rat. Journal of Comparative Neurology, 151, 121-158.

Pfaff, D.W. and Sakuma, Y. (1979). Deficit in the lordosis reflex of female rats caused by lesions in the ventromedial nucleus of the hypothalamus. Journal of Physiology, London, 228, 203-210.

Phoenix, C.H., Goy, R.W. and Young, W.C. (1967). Sexual behavior : general aspects. In Martini L. and Ganong W.F. (Eds.). Neuroendocrinology, Vol. II, pp. 163-196, Academic Press, New York.

Phoenix, C.H., Jensen, J.N. and Chambers, K.C. (1983). Female sexual behavior displayed by androgenized females rhesus macaques. Hormones and Behavior, 17, 146-151.

Phoenix, C.H., Chambers, K.C., Jensen, J.N. and Baughman, W. (1984). Sexual behavior of an androgenized female rhesus macaque with a surgically constructed vagina. Hormones and Behavior, 18, 393-399.

Pomerantz, S.V., Roy, M.M., Thornton, J.E. and Goy, R.W. (1985). Expression of adult female patterns of sexual behavior by male, female and pseudohermaphrodite female rhesus monkey. Biology of Reproduction, 33, 878-889.

Powers, J.B., Fields, R.B. and Winans, S.S. (1979). Olfactory and vomeronasal system participation in male hamsters. Attraction to female vaginal secretions. Physiology and Behavior, 22, 77-84.

Price, D. and Williams-Ashman, H.G. (1961). The accessory reproductive glands of mammals. In Young W.C. (Ed.). Sex and Internal Secretions, Vol. 1, pp. 366-448, Wilkins, Baltimore.

Reddy, V.V.R., Naftolin, F. and Ryan, K.J. (1974). Conversion of androstenedione to estrone by neural tissues from fetal and neonatal rats. Endocrinology, 94, 117-121.

Redmond, A.F., Albrecht, E.D. and Pepe, G.J. (1986). Testosterone production by collagenase dispersed cells from Baboon testis. Biology of Reproduction, 35, 372-376.

Renaud, L.P. and Martin, J.B. (1975). Electrophysiological studies of connections of hypothalamic ventromedial nucleus neurons in the rat : evidence for a role in neuroendocrine regulation. Brain Research, 93, 145-151.

Reyes, F.I., Boroditsky, R.S., Winter, J.S.D. and Fairman, C. (1974). Studies on human sexual development. II. Fetal and maternal serum gonadotropin and sex steroid concentration. Journal of Clinical and Endocrinological Metabolism, 38, 612-617.

Roffi, J., Chami, F., Corbier, P. and Edwards, D.A. (1987). Testicular hormones during the first few hours after birth and the tendency of adult male rats to mount receptive females. Physiology and Behavior, 39, 625-628.

Samama, B. and Aron, C. (1989). Changes in estrogen receptors in the mediobasal hypothalamus mediate the facilitory effects exerted by the male's olfactory cues and progesterone on feminine behavior in the male rat. The Journal of Steroid Biochemistry, 32, 525-529.

Schaeffer, Ch. and Aron, C. (1980). Environnement olfactif et comportement sexuel hétérotypique chez le rat mâle castré. Comptes Rendus de l'Académie des Sciences (Paris), Série D, 290, 485-487.

Schaeffer, Ch. and Aron, C. (1981). Studies on feminine behavior in the male rat : influence of olfactory stimuli. Hormones and Behavior, 15, 377-385.

Schaeffer, Ch. and Aron, C. (1982). Facilitory effects of male urine on feminine behavior in the male rat : androgen dependency. Physiology and Behavior, 29, 677-680.

Schaeffer, Ch., Al Satli, M., Kelche, Ch. and Aron, C. (1982). Olfactory environment and lordosis behaviour in the female and male rat. In Breipohl W. (Ed.). Olfactory and Endocrine Regulation, pp. 115-126, IRL Press Ltd. London.

Schaeffer, Ch., Chabli, A. and Aron, C. (1986a). Endogenous progesterone and lordosis behavior in male rats given estrogen alone. The Journal of Steroïd Biochemistry, 25, 99-102.

Schaeffer, Ch., Roos, J. and Aron, C. (1986b). Accessory olfactory bulb lesions and lordosis behavior in the male rat feminized with ovarian hormones. Hormones and Behavior, 20, 118-127.

Schreiner, L. and Kling, A. (1953). Behavioral changes following rhinencephalic injury in cats. Journal of Neurophysiology, 16, 643-659.

Siiteri, P.K. and Wilson, J.D. (1974). Testosterone formation and metabolism during male sexual differentiation in the human embryo. Journal of Clinical and Endocrinological Metabolism, 38, 113-125.

Singer, A.G. (1976). Dimethyldisulphide : an attractant pheromone in hamster vaginal secretion. Science, 191, 948-950.

Singer, A.G., Macrides, F., Clancy, A.N. and Agosta, W.C. (1986). Purification and analysis of a proteinaceous aphrodisiac pheromone from hamster vaginal discharge. Journal of Biology and Chemistry, 261, 13323-13326.

Smale, L., Nelson, R.J. and Zucker, I. (1985). Neuroendocrine responsiveness to oestradiol and male urine in neonatally androgenized prairie voles (Microtus ochrogaster). Journal of Reproduction and Fertility, 74, 491-496.

Södersten, P. (1973). Estrogen activated sexual behavior in male rats. Hormones and Behavior, 4, 247-256.

Södersten, P. and Larsson, K. (1974). Lordosis behavior in castrated male rats treated with estradiol benzoate or testosterone propionate in combination with an estrogen antagonist MER-25, and in intact male rats. Hormones and Behavior, 5, 13-18.

Södersten, P., De Jong, F.H., Vreeburg, J.T.M. and Baum, M.J. (1974). Lordosis behavior in intact male rats : absence of correlation with mounting behavior or testicular secretion of estradiol 17β and testosterone. Physiology and Behavior, 13, 803-808.

Stone, C.P. (1924). A note on "feminine" behavior in adult male rats. American Journal of Physiology, 68, 39-41.

Stumpf, W.E. (1970). Estrogen-neurons and estrogen-neuron system in the periventricular brain. American Journal of Anatomy, 129, 207-218.

Thornton, J. and Goy, R.W. (1986). Female typical sexual behavior of rhesus and feminization by androgens given prenatally. Hormones and Behavior, 20, 129-147.

Thornton, J., Wallen, K. and Goy, R.W. (1987). Lordosis behavior in males of two inbred strains of guinea pig. Physiology and Behavior, 40, 703-709.

Vague, J. and Favier, G. (1977). Hormones sexuelles et homo-sexualité. In Hormones et Sexualité. Problèmes Actuels en Endocrinologie et en Nutrition, série 21, 197-217, Expansion Scientifique Française.

Van de Poll, N.E. and Van Dis, H. (1977). Hormone induced lordosis and its relation to masculine sexual activity in male rats. Hormones and Behavior, 8, 1-7.

Ward, I.L. (1972). Prenatal stress feminizes and demasculinizes the behavior of males. Science, 175, 82-84.

Ward, I.L. and Weisz, J. (1980). Maternal stress alters plasma testosterone in fetal males. Science, 207, 328-329.

Weisz, J. and Ward, I.L. (1980). Plasma testosterone and progesterone titers in pregnant rats, their male and female fetuses and neonatal offspring. Endocrinology, 106, 306-316.

Weisz, J., Brown, B.L. and Ward, I.L. (1982). Maternal stress decreases steroid aromatase activity in brains of male and female rat fetuses. Neuroendocrinology, 35, 374-379.

Whalen, R.E. (1968). Differentiation of the neural mechanisms which control gonadotropin secretion and sexual behavior. In Diamond M. (Ed.). Perspectives in Reproduction and Sexual Behavior, pp. 303-340, Indiana Univ. Press, London.

Whalen, R.E. and Olsen, K.L. (1981). Role of aromatization in sexual differentiation : effects of prenatal ATD treatment and neonatal castration. Hormones and Behavior, 15, 107-122.

Whalen, R.E., Gladue, B.A. and Olsen, K.L. (1986). Lordotic behavior in male rats : genetic and hormonal regulation of sexual differentiation. Hormones and Behavior, 20, 73-82.

Wiegand, S.J., Watson, R.E. and Hoffman, J.E. (1986). A sexually dimorphic opiate distribution in the preoptic area of the rat : activational effects of gonadal steroids. Biology of Reproduction, 34, Suppl. 1, 214.

Yamanouchi, K. and Arai, Y. (1983). Forebrain and lower brainstem participation in facilitory and inhibitory regulation of the display of lordosis behavior in female rats. Physiology and Behavior, 30, 155-159.

Yamanouchi, K. and Arai, Y. (1985). Presence of a neural mechanism for the expression of female sexual behaviors in the male rat brain. Neuroendocrinology, 40, 393-395.

Zaborszky, L. (1982). Afferent connections of the basal medial hypothalamus. Advances in Embryology and Cellular Biology, 69, Springer Verlag, Berlin.

Zouhar, R.L. and De Groot, J. (1963). Effects of limbic brain lesions on aspects of reproduction in female rats. Anatomical Records, 145, 358 (Abstr.).

6 The Development of Sexuality and Eroticism in Human Kind

John Money *

Sexuality in animals is usually equated with, and referred to as reproductive behaviour or reproductive biology.

Reproduction is more respectable than sex. Except on documents requiring name, sex and age, sex is still a dirty word among many scientists as well as among non-scientists of the new right, and sexology as science is demoted as being bawdy rather than scholarship. Reproductive biologists and sexologists seldom attend the same meetings or publish together in the same journals.

Sexuality in human beings may also be equated with reproductive behaviour and biology, in which case it has to do chiefly with fertility and infertility, cyclicity, gestation and delivery and is evasive with respect to coition, carnal lust, and pair-bonded love. Human reproductive biology cannot, however, be divorced from love, carnal pleasures and coition, except maybe in such a test case as that of donor insemination - and even in that case, imagery may fill the void. In clinical usage, sexuality is the term that, in recent years, has been used to reunite reproduction as respectable science with carnal passion as suspect science. For centuries the church defined passion as sin, even for husband and wife (Boswell, 1980 ; Bullough, 1976) and to a residual extent, this definition still permeates the lax and society. Even today, the term sexuality as used in science excludes much of what is conveyed by the companion term, erotic. Sexuality has as its etymological root the Latin verb meaning to cut or divide - into male and female. Thus, sexuality tends not to be as wide-ranging as eroticism, which has as its etymological root, Eros, love. Eroticism embraces sexual union, but much more as well, especially in imagery including verbal ideational imagery, and fantasy. In this review, the term "imagery" always includes verbal

* Department of Psychiatry and Behavioural Sciences and Department of Pediatrics, The Johns Hopkins University and Hospital, Baltimore, Maryland, 21205 USA.

ideational imagery as well as pictorial, tactual, or any other imagery generated through sensory perception directly or retrieved from memory. For the church, sexual passion in marriage was sin enough, but eroticism ranged far beyond the constraints of marriage and so was even more sinful. Eroticism included all the unconventional expressions of sex, including what are known today as the paraphilias. Formerly they were the perversions. Some of them, particularly the fetishes, may exist independently of a partner.

There is a need for a term that signifies both the sexual and the erotic as a unity. For lack of another, erotosexual is that term. It ensures a unity between that which takes place between the ears, and that which takes place between the groins. Developmentally, the two take place as one.

REDUCTIONISM VS. MULTIVARIATE COMPLEXITY

The principle of multivariate sequential determinism is the ultimate and absolutely imperative foundation of any trustworthy theory of the development of human sexuality. Among contemporary theorists, this principle is violated more often than it is obeyed. With ontogenetic single-mindedness of purpose, people all too often follow a reductionist dogma. Theoretically, they reduce the origins and development of human sexuality to a single and usually abstrusely defined determinant which typically belongs on one side or the other of the obsolete nature nurture fence. Foolishly, they juxtapose biology against the socioculturally acquired or learned, unmindful of the fact that there is a biology of learning and memory, albeit mostly as yet undiscovered. Like the heredity environment protagonists, they wrongly equate the biological with the fixed and preordained, and the sociocultural with the unfixed and optional. By implication, the preordained is unmodifiable, and the arbitrary modifiable. Herein lurks another implication, a covertly political one. Scientists of the status quo favour a reductionist dogma of the biological unmodifiability of anything in men's and women's sexuality. Scientists of change favour another reductionist dogma, that of the sociocultural and environmental modifiability of everything in men's and women's sexuality.

In the prestige hierarchy of today's science, the biological sciences rate higher than the social sciences with respect to human sexuality ; therefore, there is a differential attraction among many scientists toward reductionist explanations derived from what is traditionally classified as biology, for example, genetics and endocrinology. Sociocultural explanations are dismissed as non-biological. Quite to the contrary, sociocultural determinants of the development of human erotosexuality are neurochemically mediated through the brain and its peripheral nervous system. Mediation is by way of the transmission of sound and light signals that enter the system by way of the ears and eyes, and chemical, pressure, and temperature signals that enter the nose, mouth, and skin. All are then transmitted as chemical and electrical signals to the brain. There should be no surprise here, for there is, in the development of human beings, an exact parallel with native language. A neonate must be human, and must have a healthy brain, in order to develop a native language. Without the stimulus input, normally auditory, from others who already use a language, the baby will not, however, develop a native language. Sexuality parallels language. A neonate needs to have been born with a healthy human brain in order to develop a human erotosexuality as male or female. The end product is not, however, fully preordained at birth. Its development requires the stimulus input of others who already differentiate and define all the manifestations of erotosexuality as male or female.

The theoretical temptation to neglect sociocultural input into the brain as a component determinant of human erotosexuality is attributable, in part to ethical restriction on human experimenta-tion, and in part to the fact that pertinent investigative techniques that are ethically acceptable have not yet been discovered. In addition, there is also the extremely influential fact that the preponderance of laboratory research is done on four-legged rodents and other subprimate species. In these species, sexuality and eroticism are governed under a hormonal dictatorship far more rigid than is the case in the primate species with their uniquely hypertrophied cerebral cortices (Beach, 1948). In human beings it need scarcely be said that erotosexualism exists as much between the ears, in the cerebral cortex, as between the groins, in the genitalia. The brain is the organ where erotosexual imagery is learned and remembered, and from which it is retrieved and communicated behaviourally and in words. Other species, no matter how they experience erotosexual imagery, cannot communicate it linguistically. Science lacks the technical expertise, as yet, to read it directly from their brains. Thus it is not possible for a non-human species to serve as an animal model for the complete experimental study of sexuality and eroticism as it applies to human beings. Animal models are, to be sure, indispensible. Yet, erotosexual theory derived from animal models alone will be insufficient to parallel and explain human erotosexualism. To illustrate : there is no human counterpart of Cnemidophorus uniparens (Crews & Fitzgerald, 1980) a parthe-nogenic species of lizard in which, though there are no males, a female is better able to mature and lay her eggs if first she is mounted by another, non-ovulatory female who goes through the motions of cloacal copulation typical of the male in other, closely-related species.

As applied to selected animal models, a reductionist theory in which hormones are cause, and behaviour is effect may suffice, but in human beings more is needed to explain erotosexual masculinity or femininity and also to explain the non-erotic aspects of gender identity and role (G-I/R).

5α - REDUCTASE

A currently popular reductionist hypothesis argued ad
absurdum is that of Imperato-McGinley and coworkers (1974, 1976,
1979) regarding the role of 5α-reductase as a determinant of
human "male sex drive" and "male gender-identity". The hypothesis
is based on a three-generation pedigree of 38 hermaphroditic
individuals born to 23 interrelated families in two inbred mountain
villages in the Dominican Republic. All have the syndrome of 46,
XY, 5α-reductase deficiency hermaphroditism. Without 5α-re-
ductase, testosterone is not converted to dihydrotestosterone.
Without dihydrotestosterone in foetal life, it is postulated, the
external genitalia of a gonadal male fail to differentiate
completely as male, and the baby is born with genitalia that
appear more female than male. At the time of puberty,
5α-reductase deficiency does not bring about any further
feminization, but is responsible for a mild to moderate demasculinization, namely, impairment of secondary sexual virilization.
Imperato-McGinley and co-authors (1974), ignoring this impairment, postulated that testosterone without dihydrotestosterone is
sufficient for the somatic virilization of puberty ; and further
claimed that the phallus-clitorine, hypospadiac, and bound down
with chordee - "enlarges to become a functional penis", which is,
in fact, anatomically impossible. These same authors further
claimed that testosterone without dihydrotestosterone was
sufficient to produce a "male sex drive" and "male gender
identity" at puberty. They based this claim on the fact that 19 of
the hermaphrodites had been assigned the female status at birth,
and that 16 of them later changed to live as men and allegedly
changed their gender identities.

To postulate testosterone as the cause of a male gender
identity solely on the basis of a hermaphrodite's change of gender
status in public is reductionist folly (Rubin & al., 1981). There are
other factors to be taken into account. In the case of these
Dominican hermaphrodites, they were known from birth onward to
have a birth defect of the sex organs, and it remained
uncorrected. Thus the parents could not assign a hermaphroditic
baby as either girl or boy with the same conviction as they could
be non-defective brothers and sisters. Likewise with the rearing,
there would always be knowledge of the defect ; and in fact the
hermaphroditic children were raised as "guevedoces" ("balls" at
twelve), and not entirely unambiguously as either girls or boys.
Hence there is every likelihood that those who had been assigned
as girls and were being reared with a girl's name would not
differentiate a girl's gender identity (or G-I/R), but an ambiguous
or boyish one. Then at puberty, with still no clinical intervention
to feminize the body hormonally and surgically, its totally
non-feminine appearance gave further confirmation of non-feminine status. The child obtained this confirmation both directly from

the appearance and proprioception of the body, and indirectly from the reaction of family members, villagers, and village authorities.

Male and female role stereotypes are rigidly dichotomous in a rural Latin-American culture. To be a mannish-appearing women without breasts, without menses, and without fertility is to be unmarriageable in a society where the unmarried daughter is a family and community liability. The commonsense conclusion for all concerned, the priest and other village authorities included, is to endorse and legally accept a change of gender status, provided the individual concerned does not repudiate it. Here the social and the hormonal definition of sex both work together congruently. The congruence is all the more favourable when male status begins at birth with assignment of the guevedoce as a boy, except that a man with a defective penis too small for a copulation, surgical repair notwithstanding, has a problem in perpetuity that no amount of professional euphemizing can ever euphemize away.

EROTOSEXUALISM AND G-I/R

Except for some types of hermaphrodite or intersex, people are born with the genital morphology of male or female. That morphology usually dictates the sex of their rearing. Probably the only exception occurs when a baby with a normal penis and empty scrotum is diagnosed as having the adrenogenital syndrome, and its found to be chromosomally 46,XX and gonadally ovarian. In some, though not all such cases (Money & Daléry, 1972), the penis is extirpated, the vagina opened up, and the sex is reassigned as female. Even such rare cases in which the external genitalia are surgically revised, serve to reinforce the universal assumption that the morphology of the external genitals of the neonate can be relied upon to prognosticate erotosexual development as male or female, respectively. So confident is this prognostication that folk wisdom has for generations wrongly attributed male and female erotosexualism to the inevitability of preordained instinct.

The prognostication actually encompasses more than erotosexualism. It applies also to the non-erotosexual aspects of being either male or female, masculine or feminine, for the totality of masculinity or femininity is greater than being simply masculine or feminine in the narrow erotosexual sense. The totality includes work and play, legal status, education, manners, etiquette, and grooming. It includes, indeed, all of one's very identity and role as boy or girl, man or woman, for male-female dimorphism perfuses an influence far beyond the narrow confines or the sex organs.

The need for some term to designate this totality was recognized as imperative early in the 1950s. I tangled with the problem of writing about not only the copulatory roles but also the overall masculine and/or feminine psychology and behaviour of hermaphroditic or intersexed individuals whose social and legal sex was, in many instances, discordant singly or severally with their chromosomal, gonadal, or morphologic sex at birth. The need was met by borrowing the term "gender" from its use in philology, to coin the expression "gender role", which was originally defined (Money, 1955) thus : "The term gender role is used to signify all those things that a person says or does to disclose himself or herself as having the status of boy or man, girl or woman, respectively. It includes but is not restricted to eroticism". Eventually it became necessary to divide gender identity from gender role, even though they are two sides of the same coin, because people proved incapable of conceptualizing their essential unity.

The subdivided definition first appeared in Money and Eh-rhardt (1972), as follows : "GENDER IDENTITY : the sameness, unity, and persistence of one's individuality as male or female (or ambivalent), in greater or lesser degree, especially as it is experienced in self-awareness and behaviour. Gender identity is the private experience of gender role, and gender role is the public expression of gender identity. GENDER ROLE : everything that a person says and does, to indicate to others or to the self, the degree in which one is male or female or ambivalent. It includes but is not restricted to sexual arousal and response. Gender role is the public expression of gender identity, and gender identity is the private experience of gender role".

Gender identity is not, as commonly misconstrued, a simple assertion of "I am a male" or "I am a female" and it does not exclude, as is also commonly misconstrued, sexual and erotic components.

G-I/R (gender-identity/role) is the acronym that unites the divided halves. It is defined (Money, 1980) as : "GENDER-IDENTITY/ROLE (G-I/R) : gender identity is the private expe-rience of gender role, and gender role is public manifestation of gender identity. Gender identity is the sameness, unity, and persistence of one's individuality as male, female, or ambivalent, in greater or lesser degree, especially as it is experienced in self-awareness and behaviour. Gender role is everything that a person says and does to indicate to others or to the self the degree that one is either male or female, or ambivalent ; it includes but is not restricted to sexual arousal and response".

The development of G-I/R from infancy onward always has an erotosexual component along with its non-erotosexual components. The degree of concordance among them is variable. Thus, it is possible to speak of a boy who develops a homosexual G-I/R that manifests itself only in erotosexual activity with a partner, and not in behaviour at the workplace, whereas another person with a homosexual G-I/R in erotosexual activity with a partner is also a "flaming queen" in public. It is wrong to say of a certain type of homosexual man that he has a masculine gender identity but a homosexual preference or object choice. The correct statement for such a man is that his G-I/R is masculine except in its erotosexual aspects. Counterpart statements apply to lesbianism.

There are many anomalies of the erotosexual component of G-I/R that do not involve the homosexual-heterosexual dimension. Thus one may speak of a sadistic or masochistic G-I/R, and exhibitionistic or voyeuristic G-I/R, and so on through the list of the approximately thirty paraphilic G-I/Rs.

GENETICS AND H-Y ANTIGEN

Formerly, the story of developmental human erotosexuality began with sex-determination by the XX or XY chromosomal constitutions. The newest addition to that process pertains to the H-Y antigen, the Y-chromosome-induced histocompatibility factor (Ohno, 1978 ; Wachtel, 1978), discovered in 1976. Until 1979, there appeared to be no exception to the rule that the human embryo would fail to differentiate the primitive gonadal anlagen into testes in the absence of H-Y antigen. Conversely, in the presence of H-Y antigen, testes would always differentiate. In 1979 the absoluteness of that rule was questioned by the findings of Eicher and co-authors (1979), namely, that male-to-female trans-sexuals, who are born with testes, are negative for the H-Y antigen. Conversely, female-to-male trans-sexuals, who are born with ovaries, are positive for the H-Y antigen. These findings, provided they are replicated (which is currently proving difficult), will alter the theory of the determinism and development of human erotosexualism in a way not yet ascertained.

PRENATAL HORMONES

In human beings, there is no evidence that the X or the Y chromosomes which inhabit all of the cells of the body, including those of the brain, have, simply by reason of their presence in the cells, a direct effect on a person's erotosexual status. Rather, they have an indirect and derivative influence by way of their governing the differentiation of the embryonic gonadal anlagen into either testes or ovaries which, in turn, govern the level of testosterone in the foetal blood stream. There is no evidence against this generalization, even when all of the body's cells contain a supernumerary X or Y chromosome, as in the 47,XXX, the 47,XXY (Klinefelter's), and the 47,XYY syndromes, despite the fact that in some patients with these syndromes cerebral cortical function may be pathological.

The situation is somewhat different when one sex chromosome is absent. There is no 45,Y syndrome, a foetus without an X chromosome being non-viable. The converse is the 45,X (Turner's) syndrome, or one of its mosaic or other variants. In Turner's syndrome, one of the embryonic consequences of the cytogenetic anomaly is that gonads fail to differentiate, and without gonads, there are no gonadal hormones, and without gonadal hormones the foetus always differentiates morphologically as female. Thus the erotosexual status of Turner's syndrome in adulthood, which is all but invariably feminine, cannot be attributed to the cytogenetic status alone, without implicating also the antecedent variables associated with a female morphology, female rearing, and female hormonal replacement and rehabilitation in teen-age and adulthood.

Turner's syndrome has importance in experimental ethics insofar as it is a human experiment of nature which is the equivalent of what in animal research would be enforced foetal agonadism or castration. There is one other such experiment of nature, one that affects human pedigrees, namely the androgen-insensitivity syndrome, a syndrome in which a 46,XY foetus differentiates morphologically as a female, there being a permanent intracellular incapacity of all cells in the body to utilize androgen.

These two syndromes raise a question regarding the role of maternal and placental hormones in the development of the mammalian embryo as male or female. In animal experiments, too much oestrogen or progestin given to the pregnant mother is incompatible with the maintenance of gestation. The presence of these hormones in normal gestation raises the question of why they do not interfere with the masculinization of the male-differentiating foetus. The answer to this question implicates serum α-foetoprotein. According to the most commonly cited hypothesis (Baum, 1979), which is derived chiefly from experimentation on rodents, this substance captures circulating serum oestradiol of maternal, placental, or other origin and protects the foetal brain of either sex from its influence. The foetal male brain, in need of oestradiol for its developmental masculinization, is then able to obtain it by converting its own circulating testicular testosterone to oestradiol intracellularly (Naftolin & al., 1975 ; McEwen, 1980). According to an alternative hypothesis (Döhler, 1978), α-foetoprotein captures all but the minimal quantity of maternal oestradiol required for feminization of the brain of the developing daughter foetus. This minimal quantity is also supplied to the foetus developing as a son, but its feminizing effect is completely obliterated and transformed into masculinization under the augmenting influence of additional oestradiol converted, intracellularly, from testosterone of foetal testicular origin (see below). There is little if any evidence that in primates aromatization is necessary for differentiation of the brain or activation of behaviour. In human beings the evidence is against α-foetoprotein acting as an oestrogen-binding agent. The more favoured hypothesis is that progesterone acts as a protective antivirilizing agent in the developing human female foetus.

Abramovich and Rowe (1973) demonstrated that in the human male foetus from the 12 th to 18 th week of gestation the levels of testosterone (mainly of testicular origin) and of androstenedione (mainly of adrenocortical origin) are higher than in the female foetus. Whether these hormones are, at this period of gestation, used by the brain is unknown. The same applies to the postnatal surge of testosterone in males, a surge that peaks between the second and third month of life ; then it declines until between the seventh and twelfth months it stabilizes at the low level of prepuberty (Forest & al., 1973a, b ; 1974). In newborn girls, the low prepubertal level of testosterone is attained by the second week of life.

In the early days of foetal endocrinology, a simple formula semed to apply (Money & Ehrhardt, 1972), namely, that with androgen in the proper amount, the foetus differentiates as male ; without androgen it differentiates as female. In all of its simplicity, this rule applied across species, and irrespective of chromosomal sex or of gonadal sex. It applied also irrespective of the source of androgen and of the organ system involved, including the brain and its subsequent mediation of sex-dimorphic behaviour. The constraints on this formula's applicability applied to the amount of hormone and to its timing, there being a critical developmental period for the induction of a specific androgenic effect. There was one exception to the rule, namely, that suppression of the differentiation of a uterus in a male foetus is dependent on a foetal Mullerian inhibiting substance or factor (MIF), of which the biochemical formula still remains unascertain. MIF is a non-androgenic secretion from the foetal testes.

The nice simplicity of the classification of the sex steroids, on the basis of their biochemical structure, as androgenic, oestrogenic, or progestinic, proved not always to correlate with the effects of their activity in vivo. In particular, it has been quite conclusively demonstrated that the effects of experimental injections of the primarily testicular hormone, testosterone, can in some instances be replicated by injections of the primarily ovarian hormone, oestradiol (Baum, 1979). The key to this apparent paradox lies jointly in the fact that the biosynthesis of sex steroidal hormones, in vivo, is from cholesterol to progestin to androgen to oestrogen ; and in the fact that the molecule of one steroid hormone can be taken up intracellularly and transformed to another. One such transformation is the intracellular aromatization of testosterone to the oestrogen, oestradiol. Another is the 5α-reductase metabolic reduction of testosterone to dihydro-testosterone (Baum, 1979).

It is now widely recognized that different components of the reproductive system, as widely separated anatomically as the brain and, say, the prostate, have a different prenatal history of the intracellular transformation and usage of androgen. In foetal life, differentiation of the pelvic genitalia as male is hypothesized to be dependent on intracellular 5α-reduction of testosterone to dihydrotestosterone (Walsh & al., 1974 ; Maes & al., 1979). By contrast, prenatal differentiation of brain pathways to mediate stereotypically masculine behaviour patterns in later life is hypothesized to depend, at least in part, on intracellular aromati-zation of testosterone to oestradiol (Naftolin & al., 1975). For both hypotheses, there is a growing body of mammalian and avian experimental support. Across species, their applicability remains to be ascertained. Within a given species, their applicability still needs to be spelled out in terms to amount, developmental timing, specific organs or structures involved, and synergism with other hormones, neurotransmitters, and related body chemistries. Claims to the contrary notwithstanding (Imperato-McGinley & al., 1974), there is, in the case of the human species no evidence to justify, as yet, the application of either the aromatization or the 5α-reductase hypothesis to the brain's differentiation and development of erotosexual status in prepuberty, adolescence, or adulthood as male, female or mixed.

MASCULINIZED, FEMINIZED, DEMASCULINIZED, DEFEMINIZED

Embryologically, the process whereby sexual dimorphism differentiates is either unitypic or ambitypic. The ambitypic process applies to the differentiating gonad, which first passes through a corticomedullary phase, after which the cortex or the medulla becomes vestigial - the cortex yields to the medulla when the organ becomes a testis, and vice versa for an ovary. Subsequently, the ambitypic process applies also to the internal differentiation of sexual dimorphism, which passes through a Mullerian/Wolffian duct phase after which either the Mullerian or the Wolffian ducts become vestigial - Mullerian duct vestigiation and Wolffian duct proliferation characterize male differentiation ; and vice versa, for female differentiation.

The next and external phase of dimorphic differentiation as male or female is unitypic. That is to say, there is only one set, not two, of undifferentiated precursors of the external genitalia. They are the same for both sexes, but they develop differently, though homologously, as male or as female. They may differentiate incompletely and appear hermaphroditically ambiguous, but they cannot differentiate as two coexistent sets, male and female. By contrast, it is possible for a gonad to differentiate as an ovotestis, and it is possible for both Mullerian and Wolffian structures to coexist in the same person.

The next phase of dimorphic differentiation as male or female partly overlaps with the external genital phase and partly extends beyond it into postnatal life. This is the phase that pertains to hormonally induced dimorphism of the brain and its governance of behaviour. The theoretical basis of earlier investigation, both experimental and clinical, was naively assimilated from folk tradition, supposing an absolute and mutually exclusive male-female dichotomy. Using the new terminology of this present review, differentiation of the brain was assumed to be unitypic, on the model of the external genitalia. To be masculinized was synonymous with being unifeminized (or defeminized), and to be feminized was synonymous with being unmasculinized (or demasculinized).

The weight of experimental evidence now requires a theoretical change in favour of the ambitypic model. That is to say, there are in the brain two schemas, each one having its own set of neural pathways. Developmental activation of the one does not reciprocally deactivate the other (Ward, 1972, 1977 ; Baum, 1979). Activation and deactivation are two separate, though possibly linked processes. To illustrate, it is possible for a male animal to mount a female whereas, under changed circumstances, it will get into the lordosis position and be mounted by another male (or possibly by a female). The behaviour of such an animal shows that masculinization, does not automatically signify defeminization, and vice versa. Masculinization and feminization may coexist, bisexually. The ratio of masculine to feminine is not a perfect bisexual 50:50. Any ratio may be represented, over the entire range from 100:0 to 0:100.

The bisexual ratio is partly a function of species differences and, within a species, of individual differences in the determinants of sexual differentiation. These differences are well illustrated in the comparison of hamster and rat (Money & Ehrhardt, 1972). Under experimental circumstances, the normal male hamster can readily be induced to manifest both mounting and lordosis, whereas lordosis is seldom observed in the male rat. It is possible, however, to manipulate experimentally the determinants of sexual behaviour in selected individuals of either species so as to make the hamster less bisexual and the rat more so. In the hamster, the method is to inject the neonate with testosterone, thus bringing about what must be called a degree of ultramasculinization. In the rat, the converse effect can be produced by deandrogenizing the neonate either by surgical castration or by hormonal anti-androgenization. A similar effect can be experimentally induced prenatally by subjecting the mother to constraint under intense bright light. The disruption of maternal hormone secretion thus produced has a derivative effect on the male foetus and, eventually, on its bisexual behavioural ratio (Ward & Weisz, 1980).

SEX-SHARED / THRESHOLD-DIMORPHIC RESPONSE

The influence of hormones, either prenatally or neonatally, on the bisexual ratio is best conceived in terms of a threshold effect, and hormonal alteration as a resetting of the threshold. Sexual dimorphism is a characteristic not of the response, per se, but of the threshold for its emergence under given circumstances of stimulus. The response is sex-shared but threshold-dimorphic.

In human beings, the catalogue of sex-shared/threshold dimorphic behaviour for which the threshold is determined in part by the influence of prenatal hormones on the central nervous system cannot be established by experimenting on human beings. One must rely instead on animal, predominantly primate models, and especially on human clinical models. One such model is the congenital masculinizing adrenogenital syndrome in 46,XX gonadal females, some of whom are prenatally so masculinized that they are born without clitoris and vulva but with a normal penis and an empty scrotum. At the opposite extreme is the congenital feminizing androgen-insensitivity syndrome in 46,XY gonadal males who, when androgen insensitivity is complete, are born with a clitoris and normal vulva and with no penis and scrotum. Between these extremes, there are various other syndromes in which at birth the external genitalia appear neither typically male nor female, but ambiguously hermaphroditic. There are the syndromes in which, because there is no unanimity among either professionals or the laity, some babies may be publicly declared and reared as boys and others, with the very same syndrome, as girls. These are the cases, concordant for prenatal history, but discordant for post-natal history, that constitute a gold mine of comparative information regarding the prenatal versus postnatal components of sex-shared/threshold-dimorphic behaviour (Money & Ehrhardt, 1972).

On the basis of comparative animal and clinical models, it is possible to delineate at least nine components of behaviour that qualify as sex-shared but threshold-dimorphic (Money, 1980).

First is the kinetic energy expenditure, which, in its more vigorous, outdoor, athletic manifestations is typically more readily elicited and prevalent in males than females, even before males reach the postpubertal stage of being, on the average, taller, heavier, and more muscular than females.

Second is roaming and becoming familiar with or marking the boundaries of the roaming range. Whereas pheromonal (odoriferous) marking is characteristic of some small animals, in primates, (including man) vision takes the place of smell. The secretion of marker pheromones is largely under the regulation of the male sex hormone and thus is more readily elicited in males than females. The extent of any sex difference in the threshold for visual marking in primates is still conjectural.

Third is competitive rivalry and assertiveness for a position in the dominance hierarchy of childhood, which is more readily elicited in boys than in girls. A position of dominance may be accorded an individual without fighting, or after a victory. Whereas fighting and aggressiveness per se are not sexually dimorphic, despite a widespread scientific assumption that they are, sensitivity to eliciting stimuli may or may not be. An example of the latter is retaliation against a deserter or rival in love or friendship, which is not sex-specific.

Fourth is fighting of predators in defense of the troop and its territory, which, among primates, is typically more readily elicited in males than in females.

Fifth is fighting in defense of the young which is more readily elicited in females than in males. Females are more fiercely alert and responsive to threats to their infants than, in general, are males.

Sixth is provision of a nest or safe place for the delivery, care, carrying, and suckling of the young. It is possible that this variable is associated with a greater prevalence of domestic neatness in girls than boys, as compared with the disarray that is the product of, among other things, vigorous kinetic energy expenditure.

Seventh is parentalism, exclusive of delivery and suckling. Retrieving, protecting, cuddling, rocking, and clinging to the young is more prevalent in girl's rehearsal play with dolls and playmates.

Eighth is sexual rehearsal play, in which evidence derived from monkeys is that juvenile males elicit presentation responses from females, and that juvenile females elicit mounting responses from males, more readily than the opposite. The taboo on human juvenile sexual rehearsal play and on its scientific investigation prohibits any present generalization regarding boys and girls.

Ninth is the possibility that the visual erotic image more readily elicits an initiating erotic response in males than in females, whereas the tactile stimulus more readily elicits a response in females. Here again no generalization can yet be made with confidence, because of the effects of the erotic taboo and erotic stereotyping in our society.

NEONATAL BONDING

Human postnatal erotosexual differentiation and development as male or female is not prewritten in the genetic code, nor in a hormonal or other prenatal coding process or program. The end of the story is not automatically preordained by its beginning, but is dependent also on postnatal input from the environment in which it is being written. As aforesaid, this same principle applies to the development of native language, which clearly depends on the prenatal development of a brain that is both human and unimpaired, and then on the postnatal input of exteroceptive language signals, usually through the ears.

At its onset, the postnatal phase of erotosexual differentiation is not gender-specific, but gender-shared. It begins neonatally in the bonding of the baby boy or girl to its mother. The bonding of the newborn to its mother begins as soon as the mother establishes eye and finger contact with it, and the same applies to the father (see review by Trause & al., 1977). Parent-infant bonding is essential to survival. Without it, there is a greatly increased risk of parental child neglect and abuse. With it, the baby is cuddled, fondled, kissed, stroked, and rocked, even before suckling is established. The haptic (related to touch) and kinesthetic components of infant-parent bonding are for the baby the prototype of what latter in life will become lover-lover bonding. Impaired parent-infant bonding becomes a prototype of impaired lover-lover bonding.

Erection of a baby boy's penis in utero has been demonstrated by sonogram. Postnatally, erections continue spontaneously in sleep, as they will do at least three times a night throughout childhood and into advanced healthy old age (Karacan & al., 1972, 1975). When the baby is awake and being bathed or diapered, more erections occur either spontaneously or in response to tactile or temperature stimulation. Evidence that there is a parallel phenomenon, possibly a vaginal blood-flow change, in the baby girl is uncertain. Whatever the final verdict, it is nonetheless clear that baby boys and girls experience their genitals differently from birth onward. It is what the genitals do, as well as what they look like, that engages the attention of the parent or other observer and elicits a gender-differentiated response. Depending upon whether the latter is positive or negative, it will, in turn, either reinforce or inhibit the baby's response. Either way, the effect will eventually become generalized to encompass far more than the reactivity of the genitalia. This is the interactive way that gender-differentiated behaviour becomes built up in boys and girls. It is also the way in which either positive or negative erotosexual foundations are laid and built upon. In the case of boys who are neonatally circumcised an additional complexity is added in so far as their neonatal behaviour as boys is partly dictated by parent-baby interaction with respect to the unanaesthetized trauma of surgery and the subsequent pain of urine on a raw penis (Richards & al., 1976).

FREUDIAN DOCTRINE

It scarcely needs saying that, ever since its enunciation early in this century, Freud's psychoanalytic doctrine of the sequence of oral, anal and phallic stages has dominated developmental conceptions of psychosexuality in childhood. This doctrine has, in fact, become too dogmatic. The concept of the first stage puts too much emphasis on suckling and the mouth, to the exclusion of the sensuousness of haptic skin contact and body motion, and it also overlooks the specificity of infant-parent bonding. The concept of the second, or anal stage, exaggerates the sensuous significance of elimination and the discipline of training at the expense of recognizing the maturational and cross-species significance of territorial marking with urinary pheromones, and of finding a private place to defaecate and cover the excrement.

In classic psychoanalytic doctrine, the first two phases are together defined as pregenital, as terminating at around age three, and as not being gender-differentiated. In actual fact, however, the differentiation of gender identity/role (G-I/R) is well advanced by age three, so that one may conceptualize the differentiation of a core gender identity, as do some contemporary psychoanalysts (e.g. Stoller, 1964), far in advance of the phallic phase and its Oedipus complex, castration anxiety, and penis envy.

The Oedipus complex may well have constituted the imagery of Freud's own adolescent masturbation fantasy, in which case its extension to all of humanity is untenable. In so far as the Oedipal drama is a metaphor of early childhood development, it can be conceptualized in terms of identification and reciprocation in sexual rehearsal play.

IDENTIFICATION / RECIPROCATION

The differentiation of G-I/R as boyish or girlish is con-
solidated postnatally by way of the paired principles of identification and reciprocation or complementation (Money & Ehrhardt, 1972). Identification means learning by copying or
imitating - doing or saying things the same way as some other
person who constitutes the identification model. Reciprocation
means learning by doing or saying things in such a way as to
reciprocate or complement some other person, who is the
reciprocation model. In either instance, the principle applies
across a wide age range. Thus, an identification model may be as
variable in age as a parent or an age-mate. Identification with a
same-sexed parent is axiomatic to many theories of child
development, though its exact prevalence and extent are probably
overestimated at the expense of identification with other models
especially siblings and members of the peer group. Identification
may also take place with models in absentia, for example, with
characters in books and on television. The same applies, vice
versa, to reciprocation, in which case reciprocating to a character
in print or on television takes place in fantasy rather than in
actuality.

In the ultimate analysis, gender identification and reciprocation do, of course, take place in the brain. Thus there are two
gender schemas in the brain, one typifying or stereotyping
masculinity, and the other femininity. In most human beings the
two schemas develop with well differentiated gender applicability - the identification schema as mine, and the complementation
schema as thine, for the other gender. One governs my gender,
and the other prophesies what to expect from those who belong to
thy gender.

Developmentally, the brain thus postnatally differentiates as
gender ambitypic. The potential for interchangeability between the
two schemas is inversely related to age. In rare clinical cases, as
in the syndrome of transvestism, the potential is never lost, and
the person is able to alternate convincingly between identification
and reciprocation as male and female in G-I/R, with two names,
two wardrobes, two personalities (Money, 1974) and two occupations. Fertility in humans does not alternate as it does in some
fish that are able to alternate their sex of breeding (Chan, 1970,
1977 ; Robertson, 1972 ; Zupane, 1980). When alternation involves
only the erotosexual component of G-I/R, in synchrony with
whether the partner is male or female, then the phenomenon is
that of bisexualism.

FLIRTATIOUS REHEARSAL PLAY

In human infants of nursery school age, erotosexual
identification and reciprocation can be observed in the context of

flirtatious behaviour between the sexes. A girl at this age can be very coquettish with, say, her father or other men of his generation as well as with a same-aged playmate. Correspondingly, a boy at the same age can play the escort role with his mother, playmate, or other person. By age five or thereabouts, same-aged playmates in some instances become involved in kindergarten romances or love affairs. Most such bonds are transient, but some eventuate in a full-blown adolescent love affair and ultimately, maybe, in marriage.

Flirtation and romantic pair-bonding are not the only manifestations of early sexual rehearsal play. Rehearsal includes also explicit erotosexuality involving body contact and the sex organs (Martinson, 1973). The best observed and recorded animal model of such rehearsal is the rhesus monkey (Goldfoot, 1977 ; Goldfoot & Wallen, 1978).

EROTOSEXUAL REHEARSAL PLAY IN MONKEYS

When reared as members of a troop, baby rhesus monkeys typically begin to engage in erotosexual rehearsal play at age 3 months, approximately three to three and a half years prior to puberty. At first they climb on each other in all directions, in pairs or threes or fours, indiscriminate as to the sex of the partner, and as to which sex does the mounting, and which the presenting. At this stage of differentiation, the mounting and presenting behaviour qualifies as ambitypic. Eventually it becomes more unitypic, but the sex of the partner depends partly on whether the peer group is sex-segregated or coeducational.

If there are, by experimental design, no males in the peer group, then there develops a hierarchy in which some females are more dominant and do more mounting. The converse happens in all-male peer groups, some being mounted more than they mount. In a mixed-sex group, the moves eventually become sorted out so that males predominantly mount females, and do so with the foot-clasp mount typical of mature mating - that is, with their own feet off the ground and grasping the shanks of the female, who presents in the quadrupedal position.

If totally deprived of erotosexual rehearsal play by being reared without age-similar playmates, a monkey, male or female, grows up unable to assume a position for mating, even with a gentle and experienced partner, and hence unable to reproduce its species. Less than total deprivation of erotosexual rehearsal play reduces the severity of the subsequent impairment. Thus, when male monkeys, permanently separated from their mothers at 3 to 6 months of age, were allowed half an hour of peer group play daily, about 30 per cent of them developed success at foot-clasp mounting, but its appearance was delayed by a year or more until they were 18 to 24 months of age. The remaining 70 per cent remained permanently impaired, unable to mount and breed. Even the successful 30 per cent were less proficient at copulation and breeding than animals reared in the wild.

EROTOSEXUAL REHEARSAL PLAY IN HUMANS

The erotosexual rehearsal play of monkeys needs the stimulus of playful age mates, but not, according to present evidence, an actual demonstration of other monkeys in the copulatory position. In human beings at ages 3 to 4 years, spontaneous erotosexual rehearsal play can be observed as pelvic rocking or thrusting movements as children lie side by side at nap time (Money & al., 1970). The full range of erotosexual rehearsal play at this age should have been documented and classified in those few ethnic societies that do not prohibit it, but it has not been. Likewise, there is no documentation of whether children at age 5 to 6 years in these societies spontaneously extend their erotosexual rehearsal play to include coital positioning, or whether they copy positioning from the example of others who are older, but they do engage in coital positioning play from time to time. They do so without being socially obtrusive or objectionable.

The great majority of the world's children are raised under the influence of a stringent religious taboo on sex. Their infantile erotosexual rehearsal play, whether alone or with playmates, is subject to severe, often brutal, reprisals if it is discovered. Yet the evidence from rhesus monkeys confirms that copulatory rehearsal in infancy is an age-specific precursor and absolute prerequisite of successful copulation and breeding in adulthood.

LATENCY

Erotosexual rehearsal play in monkeys begins in infancy and continues in prepuberty. There is no period of latency. Psychoanalytic doctrine notwithstanding, there is also no latency period in the erotosexual rehearsal play of human children. There is, however, a period when children assimilate the sexual taboo of our society. They collude in obeying the age-avoidancy demands of the sexual taboo, and practice erotosexual privacy, modesty, prudery and neglect. In accordance with the allosex-avoidancy demands of the taboo, they do the same with age-mates of the opposite sex.

Maximally imposed, the sexual taboo requires not only privacy, modesty, prudery and neglect, but complete suppression and eradication of erotosexual rehearsal play in prepuberty. The sanctions imposed for disobedience can be very traumatic and abusive. In addition to standard threats, beatings, and deprivations, they include bizarre assaults on the genitals with a knife or scissors, threats of amputation, or threats of infibulation with a needle and thread.

The long-term developmental effects of the prohibition, prevention, and punishment of prepubertal erotosexual rehearsal play, which in the so-called latency years includes heterosexual rehearsal play, have received negligible scientific attention. Texts on child development unanimously omit erotosexual development in prepuberty. It is the only aspect of child development that is off limits to empirical science. The subject is totally neglected in paediatric health care and in preventive paediatrics. Parents and professional persons collude in not finding out whether erotosexuality is developing healthily or pathologically in a child. To maintain ignorance is to maintain also the moral myth of innocence and the scientific myth of latency. Contradictory evidence is denied, neglected, or misconstrued - especially misconstrued as having been caught from the contagious bad influence of someone else.

PUBERTY

Proscriptions on erotosexuality in childhood notwithstanding, there are some recorded instances during prepuberty of incongruities that come into full flower only after puberty. For example, in longitudinal study, it has been demonstrated that persistent incongruity of gender role in its non-erotosexual aspects in prepuberty evolves after puberty into a healthy homosexual G-I/R (Money & Russo, 1979). For the most part, however, it is only after the hormones of puberty lower the threshold for the emergence of erotosexual behaviour and imagery (typically known only through the filter of verbal report) that the products of prepubertal development first become observed. The hormones of puberty activate or release patterns of erotosexual behaviour or imagery for which the template already exists. They do not create the template.

In vernacular parlance, one says that the sex hormones of puberty increase the sexual drive. Drive is a motivational concept, too amorphous to define operationally. In the present writing, motivation as an explanatory principle has not been used, and the principle of threshold has been substituted.

The mechanism governing the onset of puberty recently was traced by Wildt & al. (1980) to the arcuate region of the mediobasal hypothalamus and specifically to its pulsatile release of gonadotrophin-releasing hormone (GnRH). In prepubertal female rhesus monkeys, experimental stimulation of pulsatile GnRH release in pulses of 6 minutes once every hour induced hormonal puberty, complete with cycles of ovulation. To be effective, pulsatile, not tonic, release of GnRH was essential. Upon cessation of the experiment, the animals reverted to a prepubertal hormonal status and stayed that way for several months until, when aged between 24 and 30 months, they became pubertal as normally expected. Only then, apparently, was the biological clock of the mediobasal hypothalamus set to begin its pulsatile function. The governance of the setting mechanism remains, as yet, unknown. Speculatively, the pineal has been implicated.

In human beings, the setting of the biological clock of puberty ranges widely, from as early as the first year of life, at the precocious extreme, to late teenage, at the delayed extreme. In either precocity or delay, there is a discrepancy between the age of the physique and the chronological age, while the social age (which includes the erotosexual age) is anchored firmly to neither one, but is closer to the chronological age, and also close to the social age of the friendship peer group. Pubertally precocious children do not become erotosexually wild and delinquent (Money & Alexander, 1969 ; Money & Walker, 1971), and pubertally delayed teenagers are not inevitably erotosexually inert, though they tend to be neglected by their pubertally developed age-mates, except in non-romantic comradeships (Money & Clopper, 1975 ; Clopper & al., 1976 ; Money & al., 1980). Though rare, it is not unknown for an older, pubertally undeveloped youth to have begun his sex life, to have sexual intercourse, and to achieve a dry-run orgasm (Money & Alexander, 1967).

HOMOSEXUAL / BISEXUAL / HETEROSEXUAL

Hormonal investigations designed to distinguish homosexuals from heterosexuals have in general been directed more to steroidal hormones from the gonads than to peptide hormones from the pituitary, presumably because masculinity is falsely equated with androgen and femininity with oestrogen, as aforesaid. Claims to the contrary notwithstanding, the overall verdict is that the hormones of puberty do not cause one's erotosexual status to be heterosexual, homosexual, or bisexual (Parks & al., 1974 ; Jaffe & al., 1980).

More information is needed regarding erotosexual status and gonadal-pituitary-hypothalamic interaction, for the weight of evidence from animals points to prenatal hormonal brain effects far more than postnatal ones as holding the key to the homosexual-heterosexual continuum. There are two published studies in which this more sophisticated approach has been utilized. Dörner and co-authors (1975), in a study still not replicated, measured the feedback effect of an injection of conjugated oestrogens (Premarin) on the hypothalamic-pituitary governance of LH (luteinizing hormone) release from the pituitary. The subjects were men located in a venereal and skin disease clinic and were homosexual, bisexual or heterosexual. Homosexual males were said to have first a decrease in serum LH and then a rebound elevation to a higher level than before receiving Premarin. The other men did not have the rebound elevation. Normal heterosexual females have a very high rebound. Seyler & al. (1978) test-treated female-to-male trans-sexual candidates with DES (diethylstilboestrol) administered orally for a week and then gave them an injection of LHRH (luteinizing-hormone releasing hormone). Heterosexual women so treated have an elevation of pituitary LH and FSH (follicle-stimulating hormone). In the female-to-male trans-sexuals, the elevation response was weak, and close to that of heterosexual men.

Concepts do not exist without terminology, but terminology may entrap and restrict concepts. The term, homosexual, restricts the concept to sex (same sex) at the expense of love (homophilia). So it is that science has long failed to recognize that the defining characteristic of the homosexual person is not sex, but love. The complete homosexual is a person who can fall in love or pair-bond as a lover only with a person of the same genital morphology as the self. The complete heterosexual can fall in love only with a person without the same genital morphology as the self. Strictly speaking, the complete bisexual should be able to do both, though usually the bisexuality is not 50:50, but leans at least a little more to one sex than the other.

If the pair-bonding experience of falling in love is the criterion of homosexuality and heterosexuality among human beings, then it is naive to expect to distinguish one type from the other on the basis of circulating gonadal steroids, for falling in love is not sex dimorphic but sex unimorphic.

Homosexual pair-bondedness has been recorded in birds, for example, in mallard drakes (Schutz, 1965, 1967) and in female Western gulls (Hunt & Hunt, 1977). Long-term avian pair-bonding may be exclusively homosexual. In subhuman mammals, genital homosexual encounters may occur (Maple, 1977) but they are more sporadic than long-term, and mostly do not exclude heterosexual pairing as well. The animal model most closely approximating the human in this behaviour is that of the stumptail macaque (Chevalier-Skolnikoff, 1974). In this species a lesbian relationship complete with orgasm occurring in the female doing the mounting has been photographed (Goldfoot & al., 1980).

LIMERENCE

One of today's faulty gender stereotypes is that males, whether homosexual or heterosexual, lack the romantic and tender emotions of love, or at least the capacity to express them. In the extreme version of this stereotype, not only is men's sexuality equated with power and dominance over women, but insisting on copulation is equated with rape. The logical obverse of this stereotype is that women, whether lesbian or heterosexual, lack the erotic and exuberant emotions of lust, or at least the capacity to express them. In the extreme version of this stereotype, not only is women's sexuality equated with martyrdom and submission, but assenting to copulation is equated with devious and deceitful intention. Even though these stereotypes apply to some men and some women, they do not apply universally. It is the birthright of women as well as men to experience lust with gusto, and the destiny of men as well as women to experience the bliss of love, and the woe of love unrequited.

The pathological but not the healthy manifestations of either love or lust may be gender disparate in their prevalence. For example, the pathology of love which takes the form of genuine rape, more accurately named raptophilia (Money, 1986), is more prevalent in men than women. By contrast, the pathology of lust, formerly known as frigidity, but more accurately named anorgasmia, is more prevalent in women than men. In either sex, it may possibly be accompanied by erotic apathy and inertia.

Normal and healthy love and lust both are gender shared, not gender disparate. Falling in love is the same for males and for females. It is the same also regardless of age. The first big love affair is usually after puberty, but it can happen in prepuberty (Money, 1980), or during any other part of the life span, including advanced age.

Limerence is a new word without etymological roots coined by Tennov (1979) to name that state of being that exists in a person who falls in love. When limerence is mutual, a high euphoria ensues. When it is less than mutual, an agony of anticipation and hope flaws the euphoria. Limerence unrequited or abandoned becomes the syndrome of love-sickness. Liebowitz and Klein (1979) have proposed a preliminary hypothesis that this syndrome may be associated with a depletion in the brain of phenylethylamine. There may be a permanent impairment of limerence as a sequel to brain surgery for a pituitary tumour, and also as a deficit accompanying idiopathic hypopituitarism (Money & al., 1980).

The possessiveness of limerence in some people becomes irrational jealousy. Limerent jealousy is a watch-dog on duty to guard against a competitor or rival who might abduct one's partner. If it is too fierce, it may so restrict and alienate the partner it guards as to destroy the relationship. In the so-called crime of passion, it destroys the very partner, a paradox which is the ultimate in irrational self-sabotage and self-deprivation. Such paradoxical irrationality is characteristic of pathological jealousy. Limerent jealousy and the violence that may accompany it are not sex-dimorphic. They may occur in either sex, in women as well as men, even though violence is culturally stereotyped as male.

Ever since the 12th century, when the troubadours of Provence formulated Europe's new philosophy of romantic or courtly love (Locke, 1978 ; Valency, 1961), limerence between lovers has become progressively the basis of the marriage contract. Before that time, and especially among European courtiers and the nobility, the marriage contract had been an arranged one - arranged by the two families on the basis of uniting their wealth, lands, and power, and maintaining the unity of religion and race.

In troubadour philosophy, later exemplified in Shakespeare's Romeo and Juliet, the lover was forever unattainable in marriage. In fact, the limerent passion of the lovelorn, if discovered in marriage, would bring ecclesiastical punishment, for the church taught either celibacy and chastity, or a minimum of passionless copulation for a maximum of marital reproduction (Valency, 1961). Here are found the historical origins of the split between romantic love and carnal lust - above the belt in public, and below the belt in private. After eight centuries, the split still survives in the "if-you-love-me" formula that unlocks the door to copulation between unmarried companions.

Today's troubadours who are the writers of the lyrics of rock and roll and other popular songs, still sing of the invincibility of limerent love as the touchstone of eternal bliss, erotosexual and otherwise. They are the chief source of love education for the young. Sometimes they sing of love spurned and failed, but they are short on explanations as to why lovers prove to be mismatched.

LOVEMAPS

It is more precise to say not that lovers are mismatched, but that their lovemaps are mismatched. Lovemaps is a recently coined term (Money, 1986), defined as a developmental representation or template in the mind and in the brain depicting the idealized lover and love affair, and the idealized programme of sexuoerotic and genital activity projected in ideation and imagery and actually engaged in with the lover.

Males and females both have lovemaps. Thus they are not disparate on this criterion. However, their lovemaps are themselves disparate on the criterion of their ideation and content, which is typically heterosexual. In normal and healthy development, male and female lovemaps are heterosexually reciprocal.

The precursor of a lovemap may already be more disposed toward either femininity or masculinity at birth, though it is not fully programmed so early in life. Like native language, its complete development is contingent on socially originated sensory input, beginning neonatally with the skin senses, and progressively including hearing and vision. In other words, the postnatal development of a personal lovemap is contingent on social learning.

On the basis of our own cultural traditions and practices regarding sexual learning and development in childhood, lovemaps are less likely to be mismatched with respect to love above the belt than to lust below the belt. Above-the-belt erotosexualism is less negated in childhood than is lust below the belt. Developmentally, therefore, it is less subject either to inhibition or to eccentric distortion and circuitous expression. Lovers can imagine what to expect of one another at the above-the-belt stage of their encounter, and usually be correct.

Below-the-belt expectancies are different. They are much more frequently a source of lovemap mismatching. In adolescence and adulthood, below-the-belt eroticism, since it is heavily negated and unmonitored in childhood development, cannot be assumed to have developed to be uncomplicatedly heterosexual. It is quite likely to have developed paraphilically, that is with eccentric distortions and circuitous ways of circumventing the negations imposed on it. Thus the mutual expectancies of their lovemaps that two lovers project onto one another, reciprocally, may not match the actuality of what each can, in fact, live up to. Such mismatching eventually leads to disillusionment regarding erotosexual function and enjoyment. It may induce secondary symptoms. It may lead to estrangement, separation, and divorce. Then, by having a consequent deleterious effect on the developing erotosexual health of the children, it ensures that erotosexual pathology will be transmitted to yet another generation.

PROCEPTION / ACCEPTION / CONCEPTION

Males and females are similar, not disparate on the criterion that the totality of a human erotosexual experience is conveniently subdivided into three phases, each of which has its own health and pathology : proception, acception, and conception. By contrast, males and females are not similar but different on the criterion of what constitutes their respective roles in proception, acception, and conception.

The term proceptive was proposed by Rosenzweig (1973) as an antonym for contraceptive (both terms are contrasted with extra-ceptive or non-genital erotic behaviour). Beach (1976) independently recommended the term proception with a different nuance, to refer to the interplay, often highly stereotyped and species-specific, between the male and female prerequisite to intromission, and hence to conception.

In men and women, the proceptive phase is the phase of preparatory arousal. When two strangers first meet, the earliest phase of proceptive interaction is, according to the ethological studies of Eibl-Eibesfeld (Donahue, 1985) and Perper (1985), remarkably predictable across cultures. The following is a list of initial proceptive, or so-called courtship manoeuvers, in sequence : establishing eye contact ; holding the gaze ; blushing ; averting the gaze, eyelids drooping ; shyly returning the gaze ; squinting and smiling ; increasing vocal animation ; accelerating the flow and breathiness of speech ; talking more loudly and exaggerating trivialities ; laughing and being jovial ; rotating to face one another ; moving closer to one another ; wetting the lips ; adjusting sleeves or other items of clothing to reveal bare skin ; touching the other person, as if inadvertently ; mirroring each other's gestures ; and synchronizing each other's bodily movements. In addition to these observable signs, those that are subjectively experienced are increasing heart rate and breating rate, perspiring, and feeling butterflies in the stomach. Over a prolonged period of time, there may also be changes in eating, sleeping, dreaming, and fantasying.

Proception is comprised of not only behaviour, but also ideation and imagery. Image units (imagerons to coin a new term) in sequence become fantasies, usually visual or narrative, as in masturbation fantasies, coital fantasies, and wet dreams. Put into practice, erotosexual imagerons become practice units or practicons (to coin another new term). Fantasies become enacted in practices that are variously called courtship rituals, mating games, making-out, and foreplay. They are staged usually with a cast of two, possibly more, and with or without stage properties. There is no one-word generic term for these proceptive enactments.

In some species of animals and birds, the courtship ritual or mating dance of the proceptive phase is highly stereotyped and species-specific, with minimal individual variation. It is an essential prerequisite of copulation, which cannot be completed without it. In human beings, the imagery and enactment or staging of the proceptive phase of the individual's lovemap also is prerequisite to copulation. It is not phyletically so fixed and stereotyped as to preclude ontogenetic variability. Ontogenetic variability exists between people, however, and not within one person over time. In fact, each person's proceptive imagery and its enactment tends to be stable, personally stereotypic, and habitually reiterated over long periods of time.

There is some evidence for the hypothesis that, relative to women, men are proceptively more dependent on vision and the visual image to release erotosexual arousal and initiative, whereas females are comparatively more dependent on touch and the tactile image. Cross-cultural evidence, particularly from those societies which endorse an erotosexual initiative in females, will be needed to test the possible cultural relativity and validity of this hypothesis.

Should the hypothesis hold across cultures, it would offer support for an explanatory theory that begins prenatally, when androgen prepares the way for erotosexual arousal to be linked with visual imagery. The content of the imagery has its input during the early childhood period of erotosexual rehearsal play. When standard heterosexual rehearsal play is displaced or thwarted, then heterosexual proceptive imagery becomes either deficient or revised. Extensively revised and altered, the imagery becomes that of a paraphilia. The paraphilias are more prevalent in men than women. Men's paraphilias involved visual imagery more often than tactual, women's being more often tactual. Women's paraphilias also involve the subordinate role of being abducted, rather than the dominant role of abducting and being coercive. This distinction holds irrespective of whether the sexuoerotic orientation is heterosexual, homosexual, or bisexual. In other words, it is not typically subject to gender transposition.

The proceptive phase merges into the acceptive phase in which the two bodies accept or receive one another, mutually. The female is not unilaterally passive and receptive, and the male is not unilaterally active and intrusive. The intromissive penis accepts the vagina, and the entrapping vagina accepts the penis.

In the build-up to orgasm, fantasy imagery that may have been prominent in the proceptive phase, and essential to arousal, typically yields to concentration on only sensory body feeling.

The conceptive phase may or may not follow proception and acception. It is the phase of pregnancy and parenthood and is a sequel to, as well as a component of, erotosexualism.

DISORDERS OF PROCEPTION

In both sexes, proception may be subject to simple failure manifested as erotosexual apathy and inertia. Failure may be complete and total. It extends into the acceptive phase. The sex ratio, the prevalence, and the differential aetiology of such failure remain largely unkown. Only rarely can a hormonal deficit be implicated. In some instances, erotosexual apathy and inertia are limited to adolescence and subsequently are self-correcting. In youth and young adulthood, they may temporarily mask a paraphilia that eventually manifests itself. Except when a hormonal deficit can be authenticated, there is no syndrome-specific therapy.

Proception may be dependent on paraphilic imagery in fantasy, practice, or both. In the absence of such imagery, erotosexual arousal and performance then fails or is deficient. In everyday speech, the imagery of a paraphilia is eccentric to the point of being bizarre. But there are no fixed criteria of what is eccentric, either statistically in terms of prevalence, or ideologically in terms of non-conformity to an arbitrary standard. There is no graduated scale on which to measure when eccentric becomes bizarre. There are no absolute standards by which to define tolerance of either the eccentric or the bizarre. Some very bizarre paraphilia are playful and harmless, and some are extremely noxious and traumatic. The former include many of the fetishisms. The latter include erotosexual self-strangulation, suicidal masochism, lethal sadism, violent paedophilia (child molestation), assaultive rapism, and lust murderism.

The imagery that makes an erotosexual fantasy or practice paraphilic has both phylographic (species) and idiographic (personal) origins. For example, it is phylographically determined that primates have the brain and peripheral nervous pathways that govern licking the newborn to keep them clean of faeces and urine. It is not phylographically but idiographically determined that, in human cases of coprophilia and urophilia, these pathways will be recruited to subserve erotosexual arousal and the release of orgasm.

Coprophilia and urophilia are inclusion paraphilias. Inclusion paraphilia are those in which some image or practice that is not typically included in erotosexualism becomes included, and indeed imperative. There are also displacement paraphilias, of which voyeurism is an example. Looking at one's partner nude is typically included in erotosexualism. In voyeurism, looking, which must be surreptitious, is not simply included but actually displaces penovaginal pratice as the trigger of orgasm. The voyeur who has a regular partner for penovaginal coitus builds up his arousal by prior peeping on a stranger, or while having intercourse replays the imagery of peeping as a coital fantasy. Otherwise his orgasm fails to eventuate.

The prevalence of paraphilia in adolescence or later has not been ascertained. It vastly exceeds those cases that are ascertained in clinics and courts. In the United States it almost certainly runs into the hundreds of thousands. The sex ratio is disproportionate, according to all the clinical and legal evidence available for both teenagers and older adults, with males greatly outnumbering females, and also having a wider range of paraphilias represented among them. The differential aetiology has scarcely been investigated. In the neurosurgical literature (reviewed by Money & Pruce, 1977) there are a few cases, no more than a score, in which a paraphilia proved to be associated with a temporal lobe lesion, and was responsive to surgery. Some paraphilic patients have non-operable epilepsy, and others have clinically documented signs of neurological pathology. Of the vast majority of cases, it can be said only that the paraphilia is developmental in origin. As yet, no uniformity in the developmental biographies of paraphiliacs has been uncovered. The paraphilias do not appear like Athena from the head of Zeus, de novo, in adulthood. They afflict children at puberty and in adolescence, as well as later. They are not copied. They are not contagious. The most successful form of therapy combines hormonal antiandrogenic treatment with counselling therapy which, preferably, should be joint therapy for both the patient and partner (Money & al., 1975 ; Money, 1980, 1981).

DISORDERS OF ACCEPTION

Disorders of the acceptive erotosexual phase are those in which the genitals themselves malfunction. The malfunction may be hyperphilic, that is, in excess, or hypophilic, that is, deficient.

Scientifically, the hyperphilias have for the most part been neglected or, perhaps with bantering envy, jokingly written off as nymphomania, erotomania, satyriasis and Don Juanism. Behind the names, however, lies the fact that some men and women engage in some form of erotosexual activity with atypical frequency, or with many partners as couples or in groups. Rather than experiencing spaced diversity and variety, they may experience instead a compulsive repetitiousness that, like an addict, they cannot govern. For some the correct term is polyiterophilia, for they reiterate over and over again a limited and stereotyped pattern of behaviour, changing only their partners. Hyperphilia includes also such phenomena as priapism, a pathological failure to detumesce ; and excessive frequency of orgasm, hyperorgasmia, which can become almost the equivalent of an automatism. Hyperorgasmia is so frequently associated with paraphilia that it is virtually pathognomonic.

Until very recently it was believed that women were hyperphilic and abnormal if they claimed to ejaculate fluid at orgasm. New findings (Sevely & Bennett, 1978 ; Belzer, 1981 ; Addiego & al., 1981 ; Perry & Whipple, 1981) indicate that some women do secrete an ejaculatory fluid from the periurethal glands (the female homologue of the prostate) ; and that these glands engorge to form the Grafenberg spot from which orgasm can be released by either digital or penile pressure.

Recent findings have also confirmed that it is not hyperphilic for a man to be able, prior to ejaculating, to feel several sets of muscular contractions, measurable intra-anally, of the same type as occur with ejaculation (Robbins & Jensen, 1978 ; Bohlen & al., 1980). These sets of contractions correspond to multiple orgasms experienced by some women prior to the final, big climax.

The hypophilias constitute the majority of the case load of today's sex therapists. In both sexes, they include coital anin- sertia, which may carry over from proceptive fears, or may be a specific phobia of the penis or of the vagina ; genital anaes- thesia ; and coital pain or dyspareunia. The additional male hypophilias of impotence and premature ejaculation have, as their female counterparts, lubrication failure and vaginismus.

Prevalence regarding sex-ratio and age statistics for the various hyperphilias and hypophilias have not been ascertained. The hypophilias, especially impotence and premature ejaculation in males, and in females, coital aninsertia (phobia of penetration) and anorgasmia, have been conjectured to affect up to 50 per cent of the population, regardless of age, though the criteria of evaluation are not standardized. The differential aetiology ranges widely and includes : birth defects ; prenatal brain hormonal anomaly ; hormonal target-organ anomaly ; toxic substance, prescribed or unprescribed ; infection ; neoplasm ; surgical or accidental trauma ; vascular anomaly ; peripheral or cerebral neurologic anomaly and neuropsychogenic dysfunction.

Recent publications on the aetiology of impotence have shown that penile vascular disease and low penile blood pressure (lower at the end of the waking days than after lying down sleeping) are more frequent than has been traditionally believed (Michal & Prospichal, 1978). Hormonal, especially pituitary disorder, has also been shown to have been too readily overlooked (Spark & al., 1980). Hyperprolactinaemia, for example, has recently been implicated in some cases of impotence in men, as well as anovulation in women (Kolata, 1978 ; Merceron & al., 1977). Bromocriptine therapy lowers prolactin levels. Since prolactin is elevated in response to external stress, the differential aetiology of hyperprolactinaemia must always be established, and not be taken for granted as originating in a pituitary lesion.

The appropriate treatment of hyperphilias and hypophilias is,

of course, related to their aetiology. For hypophilias, the current
vogue is for various modes of behavioural or psychological
therapy, which together constitute the new sex therapy.

EROTOSEXUAL CYCLICITY

The example of animals that copulate only when the female is
ovulatory and in heat raises the question of whether there is a
copulatory cycle synchronous with the menstrual cycle in
primates. There has been a modest amount of human research (see
reviews by McCauley & Ehrhardt, 1976 ; Baum & al., 1977 ;
Money, 1980) directed toward the hypothesis of cyclic synchrony
between hormones and behaviour in women, and a minor amount
of research concerning male and female synchrony. A deficiency
in all of this literature is methodological. No studies have been
designed to be comprenhensively multivariate in approach. To
illustrate, it is not feasible to investigate the synchrony of human
coital frequency with the hormonal phase of the menstrual cycle
without taking into account the personal moral and religious
concepts and taboos of both partners regarding menstrual
uncleanliness, orgasm by digital or oral stimulation, contraception,
scheduled pregnancy, fear of pregnancy, and copulation as an
enforced marital and reproductive duty. It is necessary to take
into account also the partner's knowledge of ovulation and
fertility, ovulatory and menstrual cramping, premenstrual
symptoms, and the history of coital apathy, phobia, pain or
dysfunction, including anorgasmia. There exists also the possibility
of seasonal or other cycles superimposed on the menstrual cycle.
Rossi and Rossi (1977) showed that menstrual mood patterns vary
according to working days versus weekend days of work.
Englander-Golden & al. (1980) showed that self-reports on arousal
and menstrual synchrony varied according to women's awareness
of the purpose of the investigators.

With so many variables likely to influence cyclic menstrual
synchrony of both erotosexuality and the general sense of
well-being, it comes as no surprise that there is frequent lack of
consistency among published findings. Such consistency as can be
found indicates (Money, 1980) there may be a cyclic peak of
erotosexual arousal in either imagery or practice, or both, as
verbally reported, at around the time of the ovulatory peaking of
oestrogen and LH. This is the time when there is a peak in
olfactory acuity (Money & Ehrhardt, 1972) and visual acuity
(Diamond & al., 1972). Alternatively, or additionally, there may be
an erotosexual peak that coincides with the postulated hormonal
low point of the days of menstrual bleeding. Coincident with the
ovulatory peak, plasma testosterone has been found to peak at a
level somewhat higher than that of the early follicular peak that
precedes it, and the late luteal peak that follows it (Persky & al.,
1977). This finding needs replication. In women, exogenous
testosterone enhances the initiation and prevalence, and possibly
the intensity, of erotosexual arousal in imagery (experienced as

desire) and practice (Money, 1961).

The erotosexual significance of the relationship between the circulating level of testosterone and the production of testosterone intracellularly by conversion from androstenedione is not known. Androstenedione is partly of adrenocortical origin. In primates, there is an increasing body of evidence (Gray & Gorzalka, 1980) that adrenocortical, as well as ovarian, androgen is essential to erotosexual arousal and practice in females.

Women differ in their reports as to whether they experience a peaking of erotosexualism more as an intaking and receptive yearning or an outgoing initiation of interaction. There is no systematically demonstrated relationship of this difference either to hormonal levels or cyclicity.

Independently of erotosexualism, there tends to be a low point in the sense of well-being synchronized in the luteal phase with the diminishing level of oestrogen and the build-up of progesterone in the premenstruum. In recent years this phenomenon, commonly referred to as premenstrual tension, has been hyperbolized into the premenstrual syndrome (Dalton, 1964). It is not universal, but ranges all the way from non-existence to the extreme schizophreniform pathology of the periodic psychosis of puberty. This psychosis synchronizes with the waning of the gonadotropic and oestrogenic peak of ovulation and can be controlled by treatment with a progestinic hormone that suppresses the oestrogenic peak.

The dysphoric phenomena of premenstrual tension are muted in women whose ovulation and hormonal cyclicity is suppressed by the oral contraceptive pill. Whether or not erotosexual arousal is increased or diminished by the pill is individually variable. In addition, arousal varies with the brand-name variations in hormonal content of the pill, so that no overall generalization is justified (Cullberg, 1972 ; Baum & al., 1977 ; Adams & al., 1978). Premenstrual tension in some women is followed by menstrual cramping induced by an excess secretion of prostaglandin, which brings on uterine contractions (Marx, 1979). Therapy is with an antiprostaglandin medication.

In many non-primate mammals, mating synchrony between the sexes is mediated pheromonally by way of the nose. The male responds to the scent of pheromone released from the vagina of the female at the time of ovulation. There is some evidence of a minor role of pheromones in mating synchrony in subhuman primates (Herbert, 1970 ; Michael & al., 1971), but for the most part olfactory signals are secondary to visual ones. In humankind, odour has been proposed as playing a role in establishing menstrual synchrony among women (McClintock, 1971 ; Hopson, 1979 ; Russell & al., 1980) ; but systematic evidence of pheromonal synchronization of copulation has not been forthcoming (Morris & Udry, 1978). That, of course, does not rule out the possibility that olfactory responsiveness occurs sporadically or in a subgroup minority of human beings, for example, in those who are extremely enthusiastic about oral sex.

There is some preliminary evidence, still awaiting replication,

regarding male-female couple synchrony with respect to mens-
trual-cyclic temperature fluctuations (Henderson, 1976) ; and also
with respect to the minor premenstrual peak in plasma level of
testosterone (Persky & al., 1977). In both instances, if the man
fluctuates, he does so in synchrony with the woman, when they
live together as a couple.

The impact of pregnancy, especially frequent pregnancy, on
the erotosexual synchrony of a couple has not been systematically
investigated. The pros and cons of coitus during gestation vary
according to folk belief or professional doctrine. Lack of
systematic investigation also characterizes the erotosexual
synchrony of a couple during lactation.

EROTOSEXUAL GERONTOLOGY

Inconsistency characterizes research data regarding not only
the synchrony of erotosexuality with the menstrual cycle, but also
- and for the same reason, namely, multivariate complexity - with
the menopause and, in both sexes, the later geriatric years. Most
variables have so far been studied singly, through lack of a
statistic designed to handle developmental multivariate determinism. Moreover, most studies have been cross-sectional and not
longitudinal, for the very obvious reason that longitudinal follow-up studies require logistical and financial guarantees that are
extremely difficult to obtain. Inevitably, therefore, there are more
questions here than there are answers, and too many answers that
are doctrinaire rather than empirically substantiated.

Historically, the relationship between hormones and geriatric
erotosexual decline began with Brown-Séquard's famed an-
nouncement in 1889 of rejuvenation at age 72 from self-injected
liquide orchitique (Olmsted, 1946). With a generous assist from the
media, the idea of rejuvenation grew into the popular
monkey-gland legend of geriatric lechery, and the widely held
clinical belief in testosterone as the remedy for impotence and
other symptoms of the so-called male climacteric (Maranon, 1929).

Laboratory data do not support the concept of universal
geriatric erotosexual decline developing synchronously with a
declining level of testosterone. Harman and Tsitouris (1980), for
example, in a cross-sectional study of different age groups of
healthy men, found no decline in plasma testosterone with age up
to the decade of the seventies, and in some men not until the
eighties. Earlier studies reviewed by Harman (1978) had similar
findings except that, after the age of 50 years, the prevalence of
men with low plasma testosterone progressively increased -
probably as a function of an increased prevalence of suboptimal
health.

The foregoing endocrine measures were not correlated with
erotosexual evaluations. The men in Harman's study, however,
belong in a longitudinal study that includes an erotosexual history
from which Clyde Martin (pers. commun) was able to demonstrate,

in men between 60 and 80 years of age, a statistically significant, but in absolute terms a small, correlation between testosterone level and the current history of sexual activity. Martin has concluded that most men, for most of their lives, function sexually at a level far below what might be predicted on the basis of hormone levels alone.

The most probable hypothesis regarding the relationship of hormonal level to erotosexual function in men, regardless of age, is that after an optimal level of testosterone has been attained, any excess is superfluous and is spilled out in the urine. There is no reasonable hypothesis regarding a direct relationship between either gonadotrophins or hypothalamic releasing factors on erotosexuality, but there is growing evidence that elevated pituitary prolactin has a negative effect (see above).

Syndromes of hormonal pathology excepted, it is not possible to attribute the level of erotosexual vigour to the level of any hormone. On the basis of these longitudinal and retrospective studies of men of upper educational and socioeconomic status being followed at the National Institute of Aging, Martin (1975, 1977) has found that erotosexual vigour is individually variable but not age-dependent. The earlier and more vigorous the erotosexuality of youth, the later and more vigorous the erotosexuality of aging.

The developmental changes of erotosexual maturity and aging have not been systematically classified in either sex. In the male they appear to be a lessening of the frequency of erection and ejaculation ; a lengthening of the time to achieve full erection ; an increasing frequency of obtaining an incomplete erection or of reverting to one ; a lengthening of the interval between erections and/or orgasms ; an increasing prevalence of prolonged staying power before ejaculating ; an increasing likelihood of either non-ejaculatory copulation or masturbation, with or without a subjective feeling of climax ; a maintenance of the feeling and intensity of climax and an increasing prevalence of body sensuousness or grooming without copulating. There may be an increasing dependence on the fulfillment of erotosexual imagery and fantasy rather than touch for erotosexual arousal.

Nocturnal penile tumescence is one phenomenon that, according to present evidence, remains relatively stable from young adulthood onward, with 3 or 4 episodes of erection per night and a total mean duration of 2 to 3 hours (Hursch & al., 1972 ; Karacan & al., 1975). At puberty, by contrast, the number of episodes ranges from 3 to 11, with a mean of 7 and a mean total duration of approximatively $3\frac{1}{2}$ hours (Karacan & al., 1972).

Nocturnal penile tumescence occurs chiefly during REM sleep. The corresponding phenomenon in women (Abel & al., 1979) is an intravaginal decrease in blood volume and an increase in pulse pressure. There have been no investigations of age-related changes of this female phenomenon.

The challenge of multivariate complexity is as great in

females as in males, if not greater with respect to erotosexual development at the menopause and subsequently. Some women, though by no means all, are severely afflicted with menopausal somatic symptoms and malaise that have a secondary adverse effect on erotosexuality. Menopausal hot flushes occur in synchrony with pulsatile elevations of LH triggered by pulses of hypothalamic LHRH which are synchronized with pulsatile changes of hypothalamic thermoregulation, both being hypothetically under brain catecholamine control (Tataryn & al., 1979 ; Meldrum & al., 1980).

The hormonal changes of the menopause include oestrogen depletion (Schiff & Wilson, 1978 ; Talbert, 1978). Thereafter, the fatty tissues in the labia majora shrink, the vaginal wall shrinks and becomes less elastic, vaginal lubrication lessens, and the uterus may have painful contraction (Masters & Johnson, 1970). However, many erotosexually active women do not find these changes impede their sex lives. Others find relief by treatment with low-dose replacement oestrogen (Schiff & Wilson, 1978), though there is a cost-benefit ratio that needs individual evaluation in every case.

There have been no investigations in postmenopausal women of synchronous erotosexual and hormone-level changes. Androgen of either adrenocortical or postmenopausal ovarian origin (Schiff & Wilson, 1978 ; Judd, 1980), even at a very low level might well be proved significant to postmenopausal erotosexualism, but there are at present no age-related data one way or the other.

Age-related changes in erotosexuality in women, irrespective of hormonal changes, appear on the basis of anecdotal data to include the following : a lessening frequency of lubrication ; a lengthening of the time to lubricate sufficiently for penile intromission ; an increasing frequency of partial lubrication ; a lengthening of the interval between lubrications and/or orgasms ; an increasing prevalence of prolonged staying power before climaxing ; an increasing likelihood of either genital interaction without intromission, or of masturbation, with or without a subjective feeling of climax ; maintenance of the feeling and intensity of climax ; and an increasing prevalence of body sensuousness or grooming without copulating. There may be a decreased dependence on imagery and fantasy and an increased dependence on touch for erotosexual arousal. Breast sensitivity is maintained ; so also is clitoral sensitivity, which may even be enhanced.

There is no complete explanation of why some postmenopausal women remain erotosexually active, or become more so, whereas others lose interest. Self-image is important - to feel desirable and to be desired leads to being aroused. The impediment to being desired is the shortage of available male partners who are compatible. The search may seem too difficult. The numerical disparity of men and women increases exponentially as, for each age, more men than women die, leaving more and more women alone.

For the lonely ones who do find a partner, the sexual taboo that earlier they obediently imposed on their children in some cases boomerangs. Those same children reimpose the taboo on their parents and, ever mindful of their inheritance, denounce the old people's erotosexual partnership and possible remarriage as immoral or unseemly. Thus does the sexual taboo complete its circle of negating healthy erotosexual function across the entire life span. To create conditions for healthy erotosexual development in our society is an enormous public health challenge ; but the very existence of the taboo that produces the challenge itself frustrates attempts to do something about it.

ACKNOWLEDGEMENTS

This study was supported by USPHS grants HD 00325 and HD 07111.

REFERENCES

Abel, G.G., Murphy, W.D., Becker, J.V. and Bitar, A. (1979). Women's vaginal responses during REM sleep. Journal of Sex & Marital Therapy, 5, 5-14.

Abramovich, D.R. and Rowe, P. (1973). Foetal plasma testosterone levels at mid-pregnancy and at term. Relationship to foetal sex. Journal of Endocrinology, 56, 621-622.

Adams, D.B., Gold, A.R. and Burt, A.D. (1978). Rise in female-initiated sexual activity at ovulation and its suppression by oral contraceptives. New England Journal of Medecine, 299, 1145-1150.

Addiego, F., Belzer, Jr., E.G., Comolli, J., Moger, W., Perry, J.D. and Whipple, B. (1981). Female ejaculation : a case study. Journal of Sex Research, 17, 13-21.

Baum, M.J. (1979). Differentiation of coital behavior in mammals : a comparative analysis. Neuroscience and Biobehavioral Reviews, 3, 265-284.

Baum, M.J., Everitt, B.J., Herbert, J. and Keverne, E.B. (1977). Hormonal basis of proceptivity and receptivity in female primates. Archives in Sex Behavior, 6, 173-192.

Beach, F.A. (1948). Hormones and Behavior : A Survey of Interrelationships between Endocrine Secretion and Patterns of Overt Response. Hoeber, New York.

Beach, F.A. (1976). Sexual attractivity, proceptivity, and receptivity in female mammals. Hormones and Behaviour, 7, 105-138.

Belzer, E.G. (1981). Orgasmic expulsions of women : a review and heuristic inquiry. Journal of Sex Research, 17, 1-12.

Bohlen, J., Held, J. and Sanderson, M. (1980). The male orgasm : pelvic contractions measured by anal probe. Archives in Sex Behavior, 9, 503-521.

Boswell, J. (1980). Christianity, Social Tolerance, and Homosexuality. Univ. of Chicago Press, Chicago.

Bullough, V.L. (1976). Sexual Variance in Society and History. John Wiley and Sons, New York.

Chan, S.T.H. (1970). Natural sex reversal in vertebrates. Philosophical Transactions of the Royal Society, London B, 259, 59-71.

Chan, S.T.H. (1977). Sponteneous sex reversal in fishes. In Money J. and Musaph H. (eds.). Handbook of Sexology, p. 91-105. Elsevier/North-Holland, New York.

Chevalier-Skolnikoff, S. (1974). Male-female, female-female, and male-male sexual behavior in the stumptail monkey, with special attention to the female orgasm. Archives in Sex Behavior, 3, 95-116.

Clopper, R., Adelson, J.M. and Money, J. (1976). Postpubertal psychosexual function in male hypopituitarism without hypogonadotropinism after growth hormone therapy. Journal of Sex Research, 12, 14-32.

Crews, D. and Fitzgerald, K.T. (1980). "Sexual" behavior in parthenogenetic lizards (Cnemidophorus). Proceedings of the National Academy of Science USA, 77, 499-502.

Cullberg, J. (1972). Mood changes and menstrual symptoms with different gestagen/oestrogen combinations : a double blind comparison with a placebo. Acta Psychiatrica Scandinavia, Suppl., 236, Munksgaard, Copenhagen.

Dalton, K. (1964). The Premenstrual Syndrome. Charles C., Thomas, Springfield.

Diamond, M., Diamond, A.L. and Mast, M. (1972). Visual sensitivity and sexual arousal levels during the menstrual cycle. Journal of Nervous and Mental Disease, 155, 170-176.

Döhler, K.D. (1978). Is female sexual differentiation hormone-mediated ? Trends in Neurological Sciences, 1, 138-140.

Donahue, P. (1985). The Human Animal : Who Are We ? Why Do We Behave the Way We Do ? Can We Change ? Simon and Schuster, New York.

Dörner, G., Rohde, W., Stahl, F., Krell, I. and Masius, W.G. (1975). A neuroendocrine predisposition for homosexuality in men. Archives in Sex Behavior, 4, 1-8.

Eicher, W., Spoljar, M., Cleve, H., Murken, J.D., Richter, K. and Stangel-Rutkowski, S. (1979). H-Y antigen in transsexuality. Lancet, ii, 1137-1138.

Englander-Golden, P., Chang, H.S., Whitmore, M.R. and Dienstbier, R.A. (1980). Female sexual arousal and the menstrual cycle. Journal of Human Stress, 6, 42-48.

Forest, M.G., Cathiard, A.M. and Bertrand, J.A. (1973a). Evidence of testicular activity in early infancy. Journal of Clinical Endocrinology and Metabolism, 37, 148-150.

Forest, M.G., Saez, J.M. and Bertrand, J. (1973b). Assessment of gonadal function in children. Paediatrician, 2, 102-128.

Forest, M.G., Sizonenko, P.C., Cathiard, A.M. and Bertrand, J.

(1974). Hypophysogonadal function in humans during the first year of life. I. Evidence for testicular activity in early infancy. Journal of Clinical Investigations, 53, 819-828.

Goldfoot, D.A. (1977). Sociosexual behaviors of nonhuman primates during development and maturity : social and hormonal relationships. In Schrier A.M. (ed.), Behavioral Primatology : Advances in Research Theory, Vol 1, p. 139-184. Lawrence Erlbaum, Hillsdale.

Goldfoot, D.A. and Wallen, K. (1978). Development of gender role behaviors in heterosexual and isosexual groups of infant rhesus monkeys. In Chivers D.J. and Herbert J. (eds.), Recent Advances in Primatology, Vol 1, p. 155-159. Academic Press, London.

Goldfoot, D.A., Westerborg-van Loon, H., Groeneveld, W. and Slob, A.K. (1980). Behavioral and physiological evidence of sexual climax in the female stumptailed Macaque (Macaca arctoides). Science, 208, 1477-1479.

Gray, D.S. and Gorzalka, B.B. (1980). Adrenal steroid interactions in female sexual behavior : A review. Psychoneuroendocrinology, 5, 157-175.

Harman, S.M. (1978). Clinical aspects of aging of the male reproductive system. In Schneider E.L. (ed.), The Aging Reproductive System, p. 29-58. Raven Press, New York.

Harman, S.M. and Tsitouris, P.D. (1980). Reproductive hormones in aging men. I. Measurement of sex steroids, basal luteinizing hormone and Leydig cell response to human chorionic gonadotropin. Journal of Clinical Endocrinology and Metabolism, 51, 35-40.

Henderson, M.E. (1976). Evidence for a male menstrual temperature cycle and synchrony with the female menstrual cycle. New Zealand Journal of Medicine, 84, 164.

Herbert, J. (1970). Hormones and reproductive behavior in rhesus and talapoin monkeys. Journal of Reproduction and Fertility (Suppl.), 11, 119-140.

Hopson, J.L. (1979). Scent and human behavior : olfaction or fiction ? Science News, 115, 282-283.

Hunt, G.L. Jr. and Hunt, M.W. (1977). Female-female pairing in western gulls (Larus occidentalis) in Southern California. Science, 196, 1466-1467.

Hursch, C.J., Karacan, I. and Williams, R.L. (1972). Some characteristics of nocturnal penile tumescence in early middle-aged males. Comprehensive Psychiatry, 13, 539-548.

Imperato-Mc Ginley, J. and Peterson, R.E. (1976). Male pseudohermaphroditism : The Complexities of male phenotypic development. American Journal of Medicine, 61, 251-272.

Imperato-McGinley, J., Guerrero, L., Gautier, T. and Peterson, R.E. (1974). Steroid 5α-reductase deficiency in man : An inherited form of male pseudohermaphroditism. Science, 186, 1213-1215.

Imperato-McGinley, J., Peterson, R.E., Gautier, T. and Sturla, E. (1979). Androgens and the evolution of male-gender identity among male pseudohermaphrodites with 5α-reductase deficiency. New England Journal of Medicine, 300, 1233-1237.

Jaffe, W.L., McCormack, W.M. and Vaitukaitis, J.L. (1980). Plasma hormones and the sexual preferences of men. Psychoneuroendocrinology, 5, 33-38.

Judd, H.L. (1980). Reproductive hormone metabolism in post-menopausal women. In Eskin B.A. (ed.), Menopause, Comprehensive Management p. 55-71. Masson Publishing, New York.

Karacan, I., Hursch, C.J., Williams, R.L. and Littel, R.C. (1972). Some characteristics of nocturnal penile tumescence during puberty. Pediatric Research, 6, 529-537.

Karacan, I., Williams, R.L., Thornby, J.I. and Salis, P.J. (1975). Sleep-related penile tumescence as a function of age. American Journal of Psychiatry, 132, 932-937.

Kolata, G.B. (1978). Infertility : Promising new treatments. Science, 202, 200-203.

Liebowitz, M.R. and Klein, D.F. (1979). Hysteroid dysphoria. North American Clinical Psychiatry, 2, 555-575.

Locke, F.W. (ed) (1978). Andreas Capellanus : The Art of Courtly Love. Frederick Ungar, New York.

Maes, M., Sultan, C., Zerhouni, N., Rothwell, S.W. and Migeon, C.J. (1979). Role of testosterone binding to the androgen receptor in male sexual differentiation of patients with 5α-reductase deficiency. Journal of Steroidal Biochemistry, 11, 1385-1390.

Maple, T. (1977). Unusual sexual behavior of nonhuman primates. In Money J. and Musaph H. (eds), Handbook of Sexology, p. 1167-1186. Excerpta Medica, Amsterdam.

Maranon, G. (1929). The Climacteric. Mosby, St Louis.

Martin, C.E. (1975). Marital and sexual factors in relation to age, disease, and longevity. In Wirt R.D., Winokur G. and Roff M. (eds.), Life History Research in Psychopathology, Vol 4, p. 326-347. University of Minnesota Press, Minneapolis.

Martin, C.E. (1977). Sexual activity in the ageing male. In Money J. and Musaph H. (eds.), Handbook of Sexology, p. 813-824. Excerpta Medica, Amsterdam.

Martinson, R. (1973). Infant and Child Sexuality : A Sociological Perspective. Gustavus Adolphus College, St. Peter.

Marx, J.L. (1979). Dysmenorrhea : Basic research leads to a rational therapy. Science, 205, 175-176.

Masters, W.H. and Johnson, V.E. (1970). Human Sexual Inadequacy. Little, Brown, Boston.

McCauley, E. and Ehrhardt, A.A. (1976). Female sexual response : Hormonal and behavioral interactions. Primary Care, 3, 455-476.

McClintock, M.K. (1971). Menstrual synchrony and suppression. Nature, 229, 244-245.

McEwen, B.S. (1980). Steroid hormones and the brain : Cellular mechanisms underlying neural and behavioral plasticity. Psychoneuroendocrinology, 5, 1-11.

Meldrum, D.R., Tataryn, I.V., Frumar, A.M., Erlik, Y., Lu, K.H. and Judd, H.L. (1980). Gonadotropins, estrogens, and adrenal steroids during the menopausal hot flash. Journal of Clinical Endocrinology and Metabolism, 50, 685-689

Merceron, R.E., Raymond, J.P., Courreges, J.P. and Klotz, H.P. (1977). Sexuality in hyperprolactinemics. Problèmes Actuels en Endocrinologie et en Nutrition, 21, 185-189.

Michael, R.P., Keverne, E.B. and Bonsall, R.W. (1971). Phero-mones : isolation of a male sex attractant from a female primate. Science, 172, 964-966.

Michal, V. and Prospichal, J. (1978). Phalloarteriography in the diagnosis of erectile impotence. World Journal of Surgery, 2, 239-248.

Money, J. (1955). Hermaphroditism, gender and precocity in hyperadrenocorticism : psychologic findings. Bulletin of the Johns Hopkins Hospital, 96, 253-264.

Money, J. (1961). The sex hormones and other variables in human eroticism. In Young W.C. (ed.), Sex and Internal Secretions, 3rd ed., p. 1383-1400, Williams and Wilkins, Baltimore.

Money, J. (1974). Two names, two wardrobes, two personalities. Journal of Homosexuality, 1, 65-70.

Money, J. (1980). Love and Love Sickness : The Science of Sex, Gender Difference, and Pair-Bonding. Johns Hopkins Univ. Press, Baltimore.

Money, J. (1981). Paraphilia and abuse-martyrdom : exhibitionism as a paradigm for reciprocal couple counseling combined with antiandrogen. Journal of Sex and Marital Therapy, 7, 115-123.

Money, J. (1986). Lovemaps : Clinical Concepts of Sexual/Erotic Health & Pathology, Paraphilia, and Gender Transposition in Childhood, Adolescence, & Maturity. New York, Irvington Publishers, 1986.

Money, J. and Alexander, D. (1967). Eroticism and sexual function in developmental anorchia and hyporchia with pubertal failure. Journal of Sex Research, 3, 31-47.

Money, J. and Alexander, D. (1969). Psychosexual development and absence of homosexuality in males with precocious puberty. Journal of Nervous and Mental Disease, 148, 111-123.

Money, J. and Clopper, R. (1975). Postpubertal psychosexual function in post-surgical male hypopituitarism. Journal of Sex Research, 11, 25-38.

Money, J. and Daléry, J. (1972). Iatrogenic homosexuality : Gender identity in seven 46,XX chromosomal females with hyperadrenocortical hermaphroditism born with a penis, three reared as boys, four reared as girls. Journal of Homosexuality, 1, 357-371.

Money, J. and Ehrhardt, A.A. (1972). Man and Woman, Boy and Girl : Differentiation and Dimorphism of Gender Identity from Conception to Maturity. John Hopkins Univ. Press, Baltimore.

Money, J. and Pruce, G. (1977). Psychomotor epilepsy and sexual function. In Money J. and Musaph H. (eds.), Handbook of Sexology, p. 969-977. Elsevier/North-Holland, New York.

Money, J. and Russo, A.J. (1979). Homosexual outcome of discordant gender identity/role in childhood : longitudinal follow-up. Journal of Pediatric Psychology, 4, 19-41.

Money, J. and Walker, P.A. (1971). Psychosexual development, maternalism, nonpromiscuity, and body image in 15 females with precocious puberty. Archives in Sex Behavior, 1, 45-60.

Money, J., Cawte, J.E., Bianchi, G.N. and Nurcombe, B. (1970). Sex training and traditions in Arnhem Land. British Journal of Medecine and Psychology, 43, 383-399.

Money, J., Wiedeking, C., Walker, P., Migeon, C., Meyer, W. and Borgaonkar, D. (1975). 47,XYY and 46,XY males with antisocial and/or sex-offending behavior : Antiandrogen therapy plus counseling. Psychoneuroendocrinology, 1, 165-178.

Money, J., Clopper, R. and Menefee, J. (1980). Psychosexual development in post-pubertal males with idiopathic panhypopituitarism. Journal of Sex Research, 16, 212-225.

Morris, N.M. and Udry, J.R. (1978). Pheromonal influences on human sexual behavior : and experimental search. Journal of Biosociological Science, 10, 147-157.

Naftolin, F., Ryan, K.J., Davies, I.J., Petro, A. and Kuhn, M. (1975). The formation and metabolism of estrogens in brain tissues. Advances in Biosciences, 15, 105-121.

Ohno, S. (1978). The role of HY antigen in primary sex determination. Journal of the American Medical Association, 239, 217-220.

Olmsted, J.M.D. (1946). Charles-Edouard Brown-Séguard : A Nineteenth Century Neurologist and Endocrinologist. Johns Hopkins Univ. Press, Baltimore.

Parks, G.A., Korth-Schutz, S., Penny, R., Hilding, R.F., Dumars, K.W., Frasier, S.D. and New, M.I. (1974). Variation in pituitary-gonadal function in adolescent male homosexuals and heterosexuals. Journal of Clinical Endocrinology and Metabolism, 39, 796-801.

Perper, T. (1985). Sex Signals : The Biology of Love. Philadelphia, ISI Press.

Perry, J.D. and Whipple, B. (1981). Pelvic muscle strength of female ejaculators : Evidence in support of a new theory of orgasm. Journal of Sex Research, 17, 22-39.

Persky, H., Lief, H.I., O'Brien, C.P., Strauss, D. and Miller, W. (1977). Reproductive hormone levels and sexual behavior of young couples during the menstrual cycle. In Gemme R. and Wheeler C.C. (eds.), Progress in Sexology : Selected Papers from the Proceedings of the 1976 International Congress of Sexology, p. 293-309. Plenum, New York.

Richards, M.P.M., Bernal, J.F. and Brackbill, Y. (1976). Early behavioral differences : Gender or circumcision ? Developmental Psychobiology, 9, 89-95.

Robbins, M.B. and Jensen, G.D. (1978). Multiple orgasm in males. Journal of Sex Research, 14, 21-26.

Robertson, D.R. (1972). Social control of sex reversal in a coral reef fish. Science, 177, 1007-1009.

Rosenzweig, S. (1973). Human sexual autonomy as an evolutionary attainment, anticipating proceptive sex choice and idiodynamic bisexuality. In Zubin J. and Money J. (eds.), Contemporary Sexual Behavior : Critical Issues in the 1970s, p. 189-230. Johns Hopkins University Press, Baltimore.

Rossi, A.S. and Rossi, P.E. (1977). Body time and social time : Mood patterns by menstrual cycle phase and day of the week. Sociological Science Research, 6, 273-308.

Rubin, R.T., Reinisch, J.M. and Haskett, R.F. (1981). Postnatal gonadal steroid effects on human sexually dimorphic behavior : a paradigm of hormone-environment interaction. Science, 211, 1318-1324.

Russell, M.J., Switz, G.M. and Thompson, K. (1980). Olfactory influences on the human menstrual cycle. Pharmacology, Biochemistry and Behavior, 13, 737-738.

Schiff, I. and Wilson, E. (1978). Clinical aspects of aging of the female reproductive system. In Schneider E.L. (ed.), The Aging Reproductive System, Vol. 4, p. 9-28. Raven Press, New York.

Schutz, F. (1965). Homosexualitat und Praegung. Eine experimentelle Untersuchung an Enten. Psychologishe Forschung, 28, 439-463.

Schutz, F. (1967). Homosexualität bei Tieren. In Homosexualität oder Politik mit dem S 175, p. 13-33. Rowohlt, Reinbek bei Hamburg.

Sevely, J.L. and Bennett, J.W. (1978). Concerning female ejaculation and the female prostate. Journal of Sex Research, 14, 1-20.

Seyler, L.E. Jr., Canalis, E., Spare, S. and Reichlin, S. (1978). Abnormal gonadotropin secretory responses to LRH in transsexual women after diethylstillbesterol priming. Journal of Clinical Endocrinology and Metabolism, 47, 176-183.

Spark, R.F., White, R.A. and Connolly, P.B. (1980). Impotence is not always psychogenic : Newer insights into hypothalamic-pituitary-gonadal dysfunction. Journal of the American Medical Association, 243, 750-755.

Stoller, R.J. (1964). A contribution to the study of gender identity. Inst. J. Psychoanal., 45, 220-226.

Talbert, G.B. (1978). Effect of aging of the ovaries and female gametes on reproductive capacity. In Schneider E.L. (ed.), The Aging Reproductive System, Vol. 4, p. 59-83. Raven Press, New York.

Tataryn, I.V., Meldrum, D.R., Lu, K.H., Frumar, A.M. and Judd, H.L. (1979). LH, FSH and skin temperature during menopausal hot flash. Journal of Clinical Endocrinology and Metabolism, 49(1), 152-154.

Tennov, D. (1979). Love and Limerence - The Experience of Being in Love. Stein and Day, New York.

Trause, M.A., Kennell, J. and Klaus, M. (1977). Parental attachment behavior. In Money J. and Musaph H. (eds.), Handbook of Sexology, p. 789-799. Elsevier/North-Holland, New York.

Valency, M. (1961). In Praise of Love : An Introduction to the Love-Poetry of the Renaissance. Macmillan, New York.

Wachtel, S.S. (1978). Genes and gender. The Sciences, May-June, p. 16-17, 32-33.

Walsh, P.C., Madden, J.D., Harrod, M.J., Goldstein, J.L., McDonald, P.C. and Wilson, J.D. (1974). Familial incomplete male pseudohermaphroditism, type 2. New England Journal of Medicine, 291, 944-949.

Ward, I.L. (1972). Prenatal stress feminizes and demasculinizes the behavior of males. Science, 175, 82-84.

Ward, I.L. (1977). Exogenous androgen activates female behavior in noncopulating, prenatally stressed male rats. Journal of Comparative Physiology and Psychology, 91, 465-471.

Ward, I.L. and Weisz, J. (1980). Maternal stress alters plasma testosterone in fetal males. Science, 207, 328-329.

Wildt, L., Marshall, G. and Knobil, E. (1980). Experimental induction of puberty in the infantile female Rhesus monkey. Science, 207, 1373-1375.

Zupang, G.H.K. (1980). Marine biology : life underwater with Hans Fricke. The German Tribune, 1 June, No. 943, 8.

7 Neuroendocrine Differentiation of Sex-Specific Gonadotrophin Secretion, Sexual Orientation and Gender Role Behaviour

Franziska Götz, Wolfgang Rohde and Günter Dörner *

Male and female animals are different in their genetic structure, their somatic features, and different in fundamental processes of life, e.g. metabolism, reproduction and information processing. These sex-specific differences are often expressed - especially with regard to reproduction - by different behavioural patterns. Whether some of these differences can be attributed to hormone-dependent processes of differentiation and maturation in the CNS, i.e. during critical stages of ontogenesis, is an important topic of research. The genetic (or gonosomal) sex is determined at the moment of unification of a male and a female germ cell. If a Y-bearing sperm cell fuses with an X-bearing ovum, testes will develop. If an X-bearing sperm cell fuses with an X-bearing ovum, ovaries will develop. Thus, the genetic sex normally determines the gonadal sex.

DIFFERENTIATION OF THE GONADS AND SECONDARY SEX ORGANS

It has been suggested that the development of a testis from the indifferent gonad depends on the presence of a gene (or a set of genes) on the Y-chromosome, and that the deciding factor is a product of this gene (or set of genes) with antigenic properties (the H-Y-antigen), which is present on the surface of germ cells of the heterogametic sex, i.e. in the Mammalia in male germ cells (Wachtel & al., 1975, 1977).

---

* Bereich Medizin (Charité) der Humboldt-Universität zu Berlin, Institut für Experimentelle Endokrinologie, DDR 104 Berlin, Schumannstrasse 20-21, Postfach 140.

The process of testicular differentiation is characterized by a rapid proliferation of medullary cords of cells which attract and surround the germ cells at the same time as a regression of cortical tissue is taking place. In the absence of the Y-chromosome a prospective ovary is formed by a much slower' proliferation of cortical elements and a regression of medullary tissue. Once the developing testis begins to function, a local diffusion of androgens takes place which is responsible for differentiating the male (or Wolffian) reproductive tract, while a further, non-androgenic testicular factor, possibly protein in nature (the Müllerian Inhibiting Factor - MIF), causes the presumptive female (or Müllerian) tract to regress. In contrast, the development of the female reproductive tract does not require a hormonal stimulus, as experiments with anti-androgens have shown (Neumann & al., 1970).

STEPS OF SEXUAL DIFFERENTIATION IN THE HUMAN

Organizational Processes

Figure 1 demonstrates the chronological sequence of the organizational process of the genetic, somatic and neuronal sex. In the ontogenesis of sex in the human (Dorner, 1980, 1988a) five steps can be distinguished :

1. The genetic or gonosomal sex is determined by the presence of an X- or Y-chromosome in the fertilizing cell.
2. The gonadal sex is differentiated under the control of sex-determining genes.
3. The somatic or genital sex is organized under the control of androgens and of MIF. The presence of these hormones results in male-type differentiation, whereas their absence results in female-type differentiation of the internal and external genitals between the 2nd and 4th month of prenatal life.
4. The neuronal sex, i.e. female-type or male-type gonadotrophin secretion, sexual orientation and gender role behaviour, are then organized by sex hormones, neurotransmitters and neuromodulators in the brain. The critical periods for sex-specific differentiation of the corresponding sex, mating and gender role centres in the brain, which begin during the 4th prenatal month, are not completely identical but they overlap. Moreover, different sex hormones are responsible for the organization of sex-specific gonadotrophin secretion, sexual orientation and gender role behaviour (Dörner & al., 1987) :
   - the "sex centres" controlling female-type or male-type gonadotrophin secretion are organized only by oestrogens.

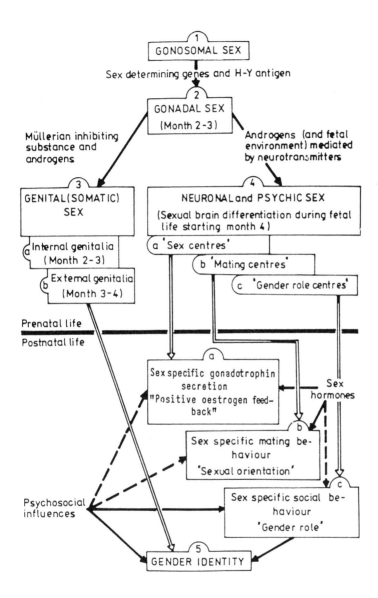

Figure 1 : Sexual differentiation in the human.

- the "mating centres" controlling sexual orientation are
  organized by oestrogens as well as androgens.
- the "gender role centres" controlling female-type or
  male-type gender role behaviour are organized only by
  androgens.

Thus, not only the absolute levels of sex hormones but also
the ratios of androgens to oestrogens are responsible for sexual
brain differentiation. Therefore, several combinations and dissoci-
ations between sex-hormone-dependent variations of gonadotrophin
secretion, sexual orientation and gender role behaviour are
possible.

5. Sexual differentiation in the human is completed by the establishment of gender identity, i.e. by a consciously experienced self-concept of being male or female. This self-concept is
   based on sex-hormone dependent differentiation of the
   somatic sex or psychic sex in prenatal life and psychosocial
   and hormonal influences in postnatal life.

Summarizing these findings, it can be assumed that during
critical developmental periods of the brain, sex hormones and/or
neurotransmitters can act as organizers of sex-specific centres,
whereas during postpubertal life activational processes in a male-
or female-type differentiated-brain largely depend also on sex hormones and/or neurotransmitters.

Terminology of Homosexuality and Trans-sexuality

In this context, a terminological problem remains to be
mentioned : is homosexuality to be considered as heterotypical
sexual behaviour related to the genetic, gonadal and genital sex or
to the self-concept of gender identity ? We prefer the heterotypical behaviour to be related to the genetic, gonadal and genital
sex, since these are the basic ontogenetic processes preceding the
differentiation of sexual orientation and gender role behaviour,
which are followed by the development of gender identity being
the last step of sexual differentiation in the human (Dörner,
1988b). Systemic hormones and/or neurotransmitters were
recognized to be organizers of the brain, e.g. of homotypical or
heterotypical behaviour. Homotypical sex hormone and/or
neurotransmitter levels (which are characteristic of the own
genetic, gonadal and genital sex) do organize a predominantly
homotypical sexual orientation and/or gender role behaviour,
whereas heterotypical sex hormone and/or neurotransmitter levels
occuring during brain organization give rise to a predominantly
heterotypical differentiation of sexual orientation and/or gender
role behaviour. In our opinion, the terms predominantly heterotypical sexual behaviour and homosexuality can and should be used

synonymously in animals and human beings.

Trans-sexuals can be divided into homo- and heterosexual trans-sexuals, while bi- and hyposexual trans-sexuals can be considered as a subgroup of heterosexual trans-sexuals (Blanchard & al., 1987). Although homosexual trans-sexuals perceive themselves as heterosexuals of the opposite sex, for the above mentioned reasons we prefer to apply for them the same terminology as for other homosexuals.

SEX-SPECIFIC GONADOTROPHIN SECRETION

Neuroendocrine Responses in Animals and Heterosexual Men

There are two criteria to distinguish a male-type differentiated "sex centre" (controlling gonadotrophin secretion) from a female-type one :

1. Males secrete gonadotrophins tonically while females have a cyclic secretion pattern. Pfeiffer (1936) was the first to discover that, independent of the genetic sex, the presence of testes during a critical perinatal period in rats resulted in acyclic gonadotrophin secretion in adulthood, whereas the absence of testes during this critical period resulted in cyclic gonadotrophin secretion in adulthood. Barraclough and Gorski (1961) then demonstrated that testosterone is the mediator of the testes for sex-specific differentiation of the brain. High androgen levels during brain differentiation gave rise to a more tonic gonadotrophin secretion in postpubertal life, whereas low androgen levels during this period gave rise to a cyclic gonadotrophin secretion in adulthood. However, the essential factor to distinguish between acyclic and cyclic gonadotrophin secretion is the presence of an ovary. Therefore, clinical and experimental research focussed more and more on the second criterion to distinguish between male-type and female-type pattern of gonadotrophin secretion.

2. The evocability of a positive oestrogen feedback effect on LH secretion, which is typical for a female-type differentiated brain. As early as in 1934 and 1937, Hohlweg and Hohlweg & al. were the first to demonstrate that a single injection of oestrogen evoked corpora lutea formation in immature female rats due to increased LH secretion. It took then some decades before a positive oestrogen action on LH secretion was also found in other mammals, including human beings (Van de Wiele & al. 1970 ; Rohde, 1981).

In 1964 the positive oestrogen feedback effect could be demonstrated to be a relatively sex-specific reaction which is dependent on the androgen level during a critical period of brain

differentiation (Dörner & Döcke, 1964). The lower the androgen
level during sexual brain differentiation, the higher was the
evocability of the positive oestrogen action on LH secretion in
adult rats (Dörner, 1976, 1980). In 1975, the evocability of the
positive oestrogen feedback was found to be additionally dependent on the oestrogen and/or androgen priming effect in adulthood
(Dörner & al., 1975a, b, c). As shown in Figure 2, an especially

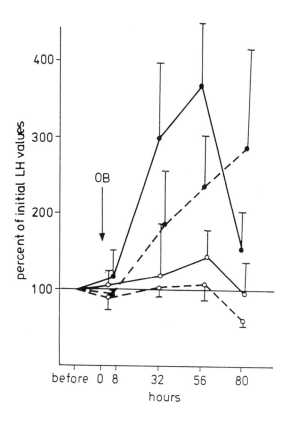

Figure 2 : Serum LH response to a subcutaneous injection of
oestradiol benzoate (OB) expressed as percent of the mean initial
LH values in postpubertally castrated and oestrogen- or androgen-
primed female and male rats (means ±SE)
       •—• castrated and oestrogen-primed female rats (n=8)
       •-• castrated and androgen-primed female rats (n=5)
       o—o castrated and oestrogen-primed male rats (n=4)
       o-o castrated and androgen-primed male rats (n=5)

high evocability of the positive oestrogen feedback effect was found after a single injection of oestrogen in castrated females following oestrogen priming. A lower and somewhat delayed but significant evocability was found in castrated females following androgen priming. On the other hand, a slight but significant evocability was also demonstrable in castrated males following oestrogen priming, but not at all in castrated males following androgen priming. In other words, a positive oestrogen action on LH secretion could only be induced by a single oestrogen injection in rats without oestrogen priming when there was a low androgen level during a critical period of brain differentiation, e.g. in females or neonatally castrated males.

Without oestrogen priming, a positive feedback effect on LH secretion could not be induced by a single oestrogen injection in pre- or postpubertally castrated male rats (Dörner & al., 1975a,b, c), hamsters (Buhl & al., 1978), pigs (Elsaesser & Parvizi, 1979) and macaque monkeys (Yamaji & al., 1971), in clear-cut contrast to pre- or postpubertally castrated females of the same species. A unique exception was found in marmoset monkeys (Hodges, 1980). In this case, the evocability of a positive oestrogen feedback action could even be demonstrated in normal intact males. Hodges (op. cit.) emphasized, however, a distinctive foetogenesis in this species with a high incidence of twinning and the presence of placental vascular anastomoses between twin foetuses.

In oestrogen-primed animals, the evocability of a positive oestrogen feedback effect on LH secretion was found to be much stronger in female as compared to male rats (Dörner & al., 1975a, b,c) as well as in female macaque monkeys compared to males (Karsch & al., 1973 ; Westfahl & al., 1984), when the LH peak values are related to the oestrogen priming and peak levels in both sexes. Steiner and associates (1976), on the other hand, observed only a sex-specific difference with regard to the negative oestrogen feedback action on LH secretion in macaque monkeys. Their animals, however, appear to have been relatively non-homogenous (body weights 5-11 kg and oestrogen administration between 3 and 36 months after castration). Nevertheless, the evocability of a slight positive oestrogen feedback action on LH secretion was demonstrated in oestrogen-primed male rats (Dörner & al.,1975a,b,c) as well as in oestrogen-primed non-human primates (Karsch & al., 1973) and oestrogen-primed men (Dörner & al., 1975a,b,c).

In this context, it should be noted that administration of oestrogen for several days can represent an oestrogen priming and positive feedback evoking effect as well. Therefore, it is not surprising at all that a relatively slight positive feedback action on LH release was readily observed in heterosexual men following several successive administrations of oestrogen (Kulin & Reiter,

1976 ; Barberino & Marinis, 1980). Such treatment - e.g. 1.5 mg/day oestradiol benzoate for 9 days - can lead to a follicle phase resembling oestradiol increase inducing a 2-3 fold LH increase (Barberino & Marinis, 1980), while in the preovulatory follicle phase of women an about 10 fold LH surge is observed. Therefore it is, however, also not surprising that a prolonged oestrogen priming in men can partly "feminize" the gonadotrophin response to oestrogen (Goh & al., 1985 ; Gooren, 1986a,b).

Neuroendocrine Response to Oestrogen in Homo-, Hetero- and Trans-sexual Men and Women

In extensive animal experiments, the outstanding importance of sex hormone levels, occuring during a critical pre- and/or early postnatal period, for sex-specific gonadotrophin secretion and/or sexual behaviour in later life was demonstrated by several authors (Pfeiffer, 1936 ; Barraclough & Gorski, 1961 ; Dantchakoff, 1938 ; Phoenix & al., 1959). Soon afterwards, the possibility of similar organizing effects of sex hormones on sexual brain differentiation in human beings was discussed (Diamond, 1965).

In 1967 it was shown that androgen deficiency in genetic male rats produced by orchidectomy during a critical perinatal period of brain differentiation gives rise to predominantly heterotypical, i.e. female-like sexual behaviour in adult males following oestrogen or even androgen treatment (Dörner, 1967, 1972, 1976, 1980). Therefore, the study of the evocability of a positive oestrogen feedback effect in intact homosexual and heterosexual men, who are both endogenously androgen primed in adulthood, appeared to be of special interest with regard to a possible aetiogenesis of "inborn homosexuality" in men due to low androgen levels during prenatal sexual brain differentiation. A positive oestrogen feedback effect, i.e. an increase of at least 3 mIU/ml LH above the initial serum levels, was found to be evocable, indeed in 13 of 21 (62 %) intact homosexual men (P<0.001) (Dörner & al., 1975a,b,c). In the group of homosexual men, a single oestrogen injection resulted in a significant decrease of the LH plasma levels which was followed by a significant increase above the mean LH initial levels. In the group of heterosexual men, on the other hand, the negative oestrogen feedback action on LH secretion was not followed by an increase above the baseline values (Figure 3).

More recently, Gooren (1986a, b) reported that the serum LH levels on day 4 after the administration of 20 mg conjugated oestrogens exceeded both initial levels in 11 of 23 homosexual men (48 %) and in 5 of 15 heterosexual men (33 %). Using the same method and the same criterion (Dörner, 1976), we had found such a neuroendocrine response to the same oestrogen dose in 14

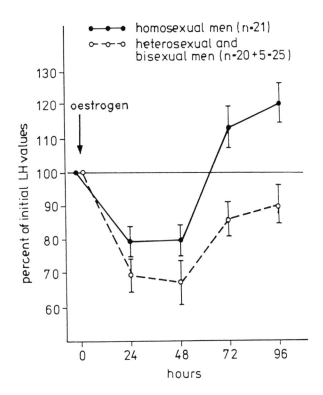

Figure 3 : Serum LH response to an intravenous injection of conjugated oestrogens (20 mg Presomen) expressed as per cent of the mean initial LH values in homosexual and hetero- or bisexual men (means ±SE). Kinsey rating 5 or 6 for homosexual and 2-4 for bisexual men.

of 21 homosexual men (57 %) and in 6 of 20 heterosexual men (30 %). Using the χ-test no significant difference was observed between the corresponding groups by both authors. Thus, one appears justified in combining the data. Subsequently, such a neuroendocrine response to oestrogen was found in 25 of 44 homosexual men (57 %), but only in 11 of 35 heterosexual men (31 %). This difference is significant (P < 0.05), although the criterion chosen by Gooren (1986a, b) does not appear to be optimal for a neuroendocrine discrimination between homo- and heterosexual men. However, in our opinion, a remarkable observation was made by Gooren, in that same paper : men with the highest LH increases 4 days after oestrogen administration showed the lowest testosterone increase 4 days after HCG administration. These data raise questions about a primary neuroendocrine abnormality as underlying the differences between homo- and heterosexuals in LH response to oestrogen.

In this context, it should be mentioned that male-type sexual differentiation of the human brain, particularly of gonadotrophin secretion, seems dependent on the responsiveness of the foetal testes to HCG in order to produce the masculinizing testosterone peak during months 4 and 5 of prenatal life (Dörner, 1980). In other words, an increased frequency of homosexuality was observed in men with Klinefelter's syndrome, who are also known to show a decreased testosterone response to HCG. In those homo- or heterosexual men in whom the LH levels on day 4 were equal to or higher than the baseline values in response to conjugated oestrogen, the testosterone response to HCG was significantly less compared to that of the men without any LH rebound (Gooren, 1986a,b). Since an increase of LH levels above the initial levels in response to oestrogen occurred significantly more often in homo-sexual than in heterosexual men (Dörner & al., 1975a,b,c ; Gladue & al., 1984), more tests comparing the testosterone response to HCG between homo- and heterosexual men should be performed in future.

Moreover, it may be noted that Pollard and Dyer (1985) observed a significant decrease in 3β-hydroxysteroid dehydro-genase (3β-HSD) activity in rat foetuses and adult animals which were exposed to stress in prenatal life ; 3β-HSD is a key enzyme for gonadotrophin-dependent testosterone synthesis in the testes. Finally, experimental and clinical findings were obtained by our group which suggest that prenatal stress is a predisposing factor for the development of predominantly heterotypical or homosexual behaviour in animals (Götz & Dörner, 1980) and human beings (Dörner & al., 1980, 1983b).

Meanwhile, our findings on the evocability of a positive oestrogen feedback action on LH secretion in homosexual men (Dörner & al., 1975a,b,c) were confirmed by Gladue & al. (1984). These authors found a positive oestrogen feedback effect, i.e. an increase in LH that exceeds mean baseline values by more than 2 standard deviations - in 9 of 14 homosexual men (64 %) compared to 0 of 17 heterosexual men (P < 0.0001) and 11 of 12 heterosexual women (89 %).

During the last five years we have studied the evocability of the positive oestrogen feedback effect in 28 male-to-female trans-sexuals. Nowadays, male-to-female trans-sexuals may be divided into two aetiologically different types, i.e. homosexual and hetero-sexual, male-to-female trans-sexuals, while asexual, hyposexual and bisexual trans-sexuals may be considered as subtypes of heterosexual trans-sexuals (Blanchard & al., 1987). A preliminary study (Dörner & al., 1983a) suggested that there are differences, indeed, between the two groups with regard to the stimulatory feedback of oestrogen on LH secretion. This study has now been completed (Rohde & al., 1986).

As can be seen in Figure 4, in homosexual plus trans-sexual men, following a decrease, a significant increase of LH secretion above the mean initial values was found after a single intravenous injection of conjugated oestrogen. A positive response - i.e. an increase of at least 3 mIU/ml LH above the initial levels - was demonstrable in 9 of 12 homosexual plus trans-sexual men (Table 1). In clear-cut contrast, in 16 hetero- or bisexual trans-sexual men, a significant decrease of serum LH levels was not followed by a significant increase above the initial levels (see Figure 3). None of the 16 hetero- or bisexual trans-sexual men reached our criterion of a positive oestrogen feedback (Table 2).

In 1976, we reported that in 3 intact, i.e. endogenously oestrogen-primed trans-sexual women with oligo- and/or hypo-menorrhoea and active homosexuality, only a slight or at best a moderate positive oestrogen feedback action on LH release was evoked (Dörner & al., 1976). A similar response had been observed in castrated and oestrogen-primed heterosexual men (Dörner & al., 1975a,b,c). By contrast, in a trans-sexual woman with eumen-orrhoea, a strong positive oestrogen feedback action on LH secretion was evocable, as well as in heterosexual women with normal gender identity (Dörner & al., 1976).

More recently, a comprehensive study of 40 female-to-male trans-sexuals was performed by Futterweit and coworkers (1986) in order to investigate the frequency of endocrine dysfunctions prior to hormonal treatment with testosterone. Two patients had lapa-roscopic evidence and 9 additional subjects had clinical evidence of polycystic ovary diseases. Plasma levels of testosterone, prolactin, LH/FSH ratio and/or dehydroepiandrosterone sulphate were significantly increased in 30 of 40 female trans-sexuals prior to testosterone treatment when compared to normal adult female controls studied in the early follicular phase of the menstrual cycle. Fourteen of the 40 patients had oligo- or amenorrhoea. We have also found oligo- or amenorrhoea in about 1/3 of our trans-sexual women (Schnabl, 1983). These data suggest that female trans-sexuals have a significantly increased incidence of endocrine dysfunctions, which may be based - at least in part - on an androgen excess acting on the brain as early as during the critical differentiation period for gonadotrophin secretion. On the other hand, normal LH responses to oestrogen should be expected and were also found in eumenorrhoic female-to-male trans-sexuals (Gooren, 1986a,b ; Dörner & al., 1976). In our opinion, eumen-orrhoea is based on a normal evocability of a positive oestrogen feedback.

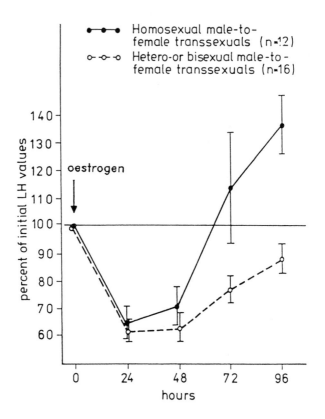

<u>Figure 4</u> : Serum LH response to an intravenous injection of conjugated oestrogens (20 mg Presomen) expressed as per cent of the mean initial LH values in homosexual and hetero- or bisexual male-to-female trans-sexuals.

Table 1 Serum LH values (mIU/ml) in 12 homosexual
male-to-female trans-sexuals before and after a
single intravenous injection of conjugated
oestrogens (20 mg Presomen).

Name	Age in years	Before oestrogen (means ± SD)	Hours after oestrogen			
			24	48	72	96
1.L.D.	22	19.0 ± 3.4	14.0	16.4	19.0	24.4
2.B.F.	26	23.6 ± 2.1	5.8	7.5	7.0	50.8
3.M.K.	29	21.4 ± 0.6	14.0	16.4	12.6	26.2
4.R.M.	38	9.5 ± 0.4	7.4	5.8	27.2	19.7
5.S.H.	35	13.6 ± 2.3	10.9	10.4	18.5	17.8
6.H.O.	47	25.8 ± 5.4	19.4	18.5	25.9	35.4
7.W.H.	25	11.4 ± 1.7	9.4	13.6	22.0	16.0
8.M.M.	20	10.0 ± 0.4	9.0	9.6	12.0	11.2
9.S.M.	21	25.0 ± 3.6	18.1	17.1	33.1	27.3
10.L.S.	19	17.3 ± 3.7	4.9	5.4	10.8	20.5
11.J.R.	21	13.1 ± 2.1	8.3	10.7	11.6	15.0
12.S.L.	25	12.6 ± 1.6	5.8	6.6	8.2	13.5
Means		16.86	10.58**	11.50 *	17.33	23.15 ***
± SD		5.96	4.81	4.75	8.32	11.05

* Significantly decreased when compared to control :
   $P < 0.01$.

** $P < 0.002$.

*** Significantly increased when compared to control :
   $P < 0.02$.

Table 2 Serum LH values (mIU/ml) in 16 hetero- or
bisexual (b) male-to-female trans-sexuals before
and after a single intravenous injection of
conjugated oestrogens (20 mg Presomen).

Name	Age in years	Before oestrogen (means ± SD)	Serum LH Values (IU/ml) Hours after oestrogen			
			24	48	72	96
1.K.H.	19	16.4 ± 1.6	9.0	8.1	6.8	14.6
2.R.W.	38	23.8 ± 1.4	16.1	15.3	15.4	6.0
3.M.K.	31	11.0 ± 0.4	4.7	7.8	7.3	9.4
4.K.D.	31	13.5 ± 0.4	10.0	10.4	11.6	11.2
5.F.C.	31	14.2 ± 1.2	8.4	7.0	7.5	12.1
6.B.D (b)	41	17.2 ± 0.0	15.2	15.0	15.2	15.8
7.T.H.	53	16.1 ± 0.1	11.2	8.8	11.6	11.0
8.K.J.	49	16.4 ± 0.6	14.4	11.2	12.4	16.8
9.B.H.	19	17.2 ± 9.8	6.6	8.2	6.7	14.4
10.S.M.	28	22.8 ± 1.5	11.2	13.7	17.5	17.5
11.B.H.	31	15.1 ± 12.8	8.6	8.1	9.8	15.7
12.G.J.	28	25.6 ± 3.7	17.0	14.8	26.7	26.0
13.B.M.	33	10.5 ± 3.5	5.2	5.4	8.3	11.8
14.L.R.	28	10.6 ± 0.9	7.9	12.4	10.0	11.6
15.L.T.	20	6.1 ± 1.1	2.2	2.1	7.8	5.6
16.S.D (b)	27	13.3 ± 1.1	10.6	9.3	10.9	14.5
Means		15.74	9.89*	9.85 *	11.59*	13.38
± SD		5.39	4.24	3.71	5.20	4.84

* Significantly decreased when compared to control : P< 0.001.

Organizational Effects of Sex Hormones in Animals

A remarkable observation was reported by Dantchakoff in 1938. Female guinea pigs, prenatally androgenized, exhibited strong male sexual behaviour when the androgen treatment was repeated in adulthood. During the last three decades, more precise quantitative studies have been carried out in order to evaluate sexual behaviour of pre- and/or postnatally androgenized rodents (Phoenix & al., 1959 ; Goy & al., 1964, 1967 ; Harris & Levine, 1962, 1965 ; Gerall & Ward, 1966 ; Dörner & Fatschel, 1970 ; Swanson & al., 1971) and monkeys (Young & al., 1964 ; Goy & al., 1972). From these findings, it can be concluded that androgens can induce a male-type differentiation of central nervous regions responsible for sexual behaviour, irrespective of the genetic sex. On the other hand, ovarian hormones appear not to be necessary for the differentiation of CNS regions that regulate female sexual behaviour (Whalen & Edwards, 1967). However, unphysiologically high oestrogen doses administered during brain differentiation are able to imitate the effects of androgens and to suppress female sexual behaviour for the entire life (Whalen & Nadler, 1963 ; Gerall, 1967).

On the other hand, several authors reported increased female sexual behaviour in male rats that had been orchidectomized immediately after birth and treated with oestrogen and progestagen during adulthood (Whalen & Edwards, 1967 ; Grady & Phoenix, 1963 ; Harris, 1964 ; Feder & Whalen, 1965). After androgen replacement in adult life, neonatally orchidectomized males showed less male sexual activity than males castrated later (Ward, 1977).

In our laboratory, parallel testing of sexual behaviour in rats of both sexes was generally carried out with both male and female partners, i.e. observing behaviour towards partners of the same and the opposite sex. This method permitted evaluation of sexual preference for or orientation towards partners of the same or opposite sex (Dörner, 1967, 1976). Testing was e.g. performed in neonatally castrated males and in perinatally androgenized females. In these tests, male rats castrated on the first day of life and treated with oestrogen or even androgen in adulthood exhibited more female-like sexual orientation i.e. "male bi- or homosexuality", while females who were androgenized in perinatal and postpubertal life displayed a complete male-like sexual orientation i.e. "female homosexuality". A complete inversion of sexual behaviour was even achieved : perinatally and post-pubertally androgenized females mounted neonatally castrated males who displayed receptive, female-like lordotic behaviour.

Organizational Effects of Sex Hormones in Humans

These experimental findings in animals combined with clinical findings in human beings speak in favour of a biological, hormone-mediated aetiogenesis of genuine homosexuality in the human. Imperato - McGinley & al. (1979) described male pseudo-hermaphrodites born with ambiguity of the external genitalia. Biochemical evaluation revealed normal testosterone levels, but a marked decrease in plasma dihydrotestosterone levels due to 5α-reductase deficiency. The decrease of dihydrotestosterone in utero resulted in incomplete masculinization of the external genitalia. Thus, the affected males were born with more female-like external genitalia and were therefore considered and raised as girls. Psychosexual orientation, however, was unequivocally male. They considered themselves males and had a libido directed towards females. Despite being reared as females, almost all of them changed gender identity at puberty. Hence, testosterone exposure in utero appears to be most important for the development of a male sex drive, male sexual orientation and even male gender identity.

Most recently, Ehrhardt & al. (1985) performed an analysis of sexual orientation in women who had been prenatally exposed to the synthetic oestrogen diethylstilboestrol. In comparison to hormone-unexposed women and even to their unexposed sisters, a variety of indicators showed statistically significant shifts towards bi- or homosexuality in the prenatally oestrogen-exposed women. In this context, it should be noted that natural and synthetic oestrogens can expert paradoxical i.e. androgen-like effects on male-type brain differentiation (Dörner & al., 1971). Several findings suggest that endogenous androgens are even aromatized in the brain to oestrogens - at least in part - for male-type brain differentiation (McEwen & al., 1977).

Stress Effects on Sexual Organizational Processes

However, prenatal psychosocial influences should also be regarded as possible aetiogenetic factors in the development of sexual deviations. Thus, Ward (1977) reported that prenatal stress in male rats demasculinized and feminized sexual behaviour potentials in adult life. Since similar findings were obtained in male rats castrated on the day of birth, we have measured the plasma testosterone levels in such prenatally stressed males, i.e. in male foetuses and newborns following maternal stress between Day 14 and 21 of gestation. The testosterone level was, in fact significantly decreased in these prenatally stressed males during the early postnatal life as compared to non-stressed control males (Stahl & al., 1978).

More recently, we observed bi- or even homosexual behaviour in prenatally stressed male rats after castration plus oestrogen treatment in adulthood, whereas prenatally non-stressed but later equally treated males displayed heterosexual behaviour (Götz & Dörner, 1980). Hence, prenatal stress can predispose to the development of bi- or even homosexual behaviour in males.

In view of these data, a retrospective study was carried out to answer the question whether stressful maternal life events occurring during pregnancy may have irreversibly affected sexual differentiation of the brain in men who were born in Germany during the stressful period of World War II. Out of about 800 homosexual males, significantly more homosexuals were born during the stressful war and early post-war period than in the years before or after World War II (Dorner & al., 1980). This finding suggested that stressful maternal life events, if occurring during pregnancy, may represent, in fact, an aetiogenetic risk factor for the development of sexual variations in the male offspring.

In addition, 100 bi- or homosexual men as well as 100 heterosexual men of similar age were asked about maternal stressful events that might have occurred during their prenatal life. Indeed, a highly significantly increased incidence of prenatal stress was found in bisexual and, in particular, in homosexual men as compared to heterosexual men (Dörner & al., 1983b). About 1/3 of the homosexual men reported exposure to severe maternal stress - such as bereavement, repudiation by the partner, rape or severe anxiety - and about an additional 1/3 to moderate maternal stress during their prenatal life. On the other hand, none of the heterosexual men was found to have been exposed to severe and less than 10 % to moderate maternal stress during their prenatal life. These data also indicate that prenatal stress may represent a risk factor for the aetiogenesis of sexual variations in later life.

In male rats, prenatal stress led to a significant decrease of plasma testosterone levels and hypothalamic norepinephrine levels of foetuses and neonates followed by hypo-, bi- or even homosexual behaviour in adulthood. Such perinatal biochemical changes and postnatal long-term behavioural changes produced by prenatal stress in male rats could be prevented by perinatal administration of testosterone and - at least in part - by prenatal administration of the norepinephrine precursor tyrosine (Dörner, 1983 ; Ohkawa & al., 1986).

In female rats, acute prenatal stress, i.e. enforced immobilization for 2 hours on day 20 of gestation, resulted in a significant decrease of β-endorphin content in the foetal pituitary associated with significant increases in levels of plasma corticosterone and androstenedione. Furthermore, daily exposure to prenatal stress between day 15 and 21 of pregnancy increased significantly male-type gender role behaviour i.e. play-fighting in prepubertal life (Figure 5) as well as heterotypical male sexual behaviour in postpubertal life (Figure 6).

These changes were combined with significantly reduced ovarian weights and some irregularities of the ovarian cycles. All these short- and long-term effects of prenatal stress could be

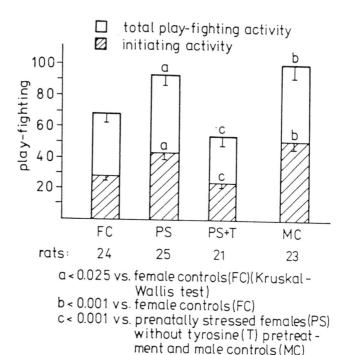

a < 0.025 vs. female controls (FC) (Kruskal-
          Wallis test)
b < 0.001 vs. female controls (FC)
c < 0.001 vs. prenatally stressed females (PS)
          without tyrosine (T) pretreat-
          ment and male controls (MC)

Figure 5 : Mean (± SE) number of 20-s sessions in which female and male controls as well as prenatally stressed females without or with tyrosine pretreatment (tyrosine methyl/ester i.p. 200 mg/kg 30 min before stress) between days 15 and 21 of gestation were observed to be engaged in playfighting (70 observation sessions per rat and day between days 26-40).

prevented, at least in part, by simultaneous administration of the norepinephrine precursor tyrosine (Dörner, 1983 ; Ohkawa & al., 1986).

In addition, a highly significant correlation was found between the plasma corticosterone levels of pregnant rats exposed to acute stress and the androstenedione levels of their female

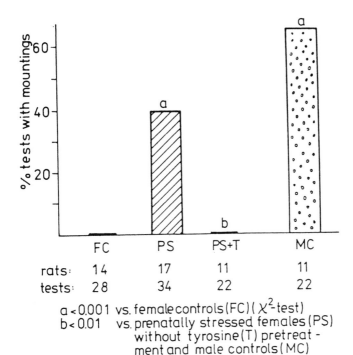

a<0.001 vs. female controls (FC) (χ^2-test)
b<0.01 vs. prenatally stressed females (PS)
          without tyrosine (T) pretreat -
          ment and male controls (MC)

Figure 6 : Male sexual behaviour (tests with mounts) in adult
female and male control rats as well as prenatally stressed
females with or without tyrosine pretreatment (see Figure 5).

foetuses. In a preliminary pilot study, a weak but significant
correlation was also observed between the salivary cortisol levels
of pregnant women exposed to slight or moderate acute stress and
the testosterone levels in the amniotic fluids of their female
foetuses.

These findings indicate that prenatal stress predisposes
females to male-type gender role behaviour and sexual orientation.
In mothers of homosexual men, undesired and illegitimate children
were often found to be the reason for stressful events during their
pregnancies (Dörner & al., 1983b). Therefore, 51 homosexual
women were also interviewed in comparison with 150 heterosexual

women whether they were born as illegitimate and/or undesired
children. As demonstrated in Figure 7, very significantly more
homosexual than heterosexual women were indeed born as
illegitimate and/or undesired children (Dörner & al., 1987).

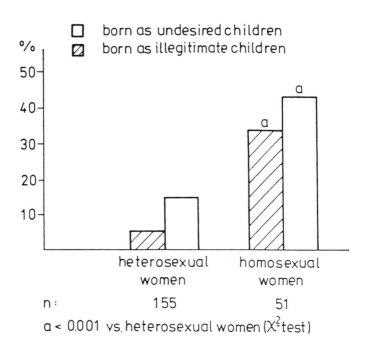

Figure 7 : Percentage of age-matched heterosexual and homosexual women who were born as illegitimate and/or undesired
children.

Thus, prenatal stress appears able to dedifferentiate
sex-specific brain organization in both sexes, i.e. reproductive
functions cannot only be inhibited in both sexes transiently by
postpubertal stress (impotence or transient amenorrhoea) but even
permanently by prenatal stress. Significantly increased 17β-
hydroxyprogesterone and androstenedione plasma levels and
decreased cortisol levels as well as significantly or even highly
significantly increased 17β-hydroxyprogesterone/cortisol and an-
drostenedione/cortisol ratios ($P < 0.002$ and < 0.001) were found
before and after ACTH stimulation in female-to-male
trans-sexuals as compared to those of female heterosexuals.
Somewhat increased 17β-hydroxyprogesterone plasma levels as

well as 17β-hydroxy-progesterone/cortisol or androstenedione/
cortisol ratios (P<0.05 or <0.02) were observed in female homosexuals without trans-sexualism as compared to female heterosexuals. Our findings suggest that a minority of non-trans-sexual
female homosexuals and, most of all, the majority of female-to-
male trans-sexuals have 21-hydroxylase deficiencies.

Effects of Neurotransmitters on Sexual Orientation in Animals

As shown in Figure 8, a conversion of female-type to male-
type sexual orientation was achieved most recently in adult rats
by administration of the dopamine agonist and serotonin antagonist
lisuride (Calne & al., 1983 ; Götz & al., 1988). In prepubertally
castrated and later androgen-treated adult females, lisuride
significantly decreased female-type lordotic behaviour towards
males and simultaneously a significant increase of male-type
mounting behaviour towards oestrous females. Thus, predominantly
homotypical ("heterosexual") behaviour in androgen-treated
females was converted to predominantly heterotypical ("homosexual") behaviour.
Furthermore, in neonatally castrated and later androgen-
treated adult males, lisuride administration resulted in a highly
significant decrease of female-type lordotic behaviour towards
vigorous males and simultaneously in a significant increase of
male-type ejaculatory behaviour towards oestrous females. Hence,
ambiguous ("bisexual") or even predominantly heterotypical
("homosexual") behaviour in neonatally castrated adult males was
converted to predominantly homotypical ("heterosexual") behaviour.
The effects of lisuride on changes of sexual orientation in
rats were even stronger than those observed after stereotaxic
lesioning of the "female mating centre" in the hypothalamic
ventromedial nuclear regions in rats (Dörner, 1969) or human
beings (Roeder & Müller, 1969 ; Müller & al., 1974). Therefore,
possible effects of lisuride on sexual orientation in the human
remain to be elucidated.

CONCLUSIONS

1. In animals and human beings the neuronal sex, i.e., female-
   type or male-type gonadotrophin secretion, sexual orientation
   and gender role behaviour, are organized by sex hormones
   and neurotransmitters in the brain. The critical periods for
   sex-specific differentiation of the corresponding sex, mating
   and gender role centres in the brain are not completely
   identical but they overlap. Moreover, different sex hormones
   are responsible for the organization of these centres.

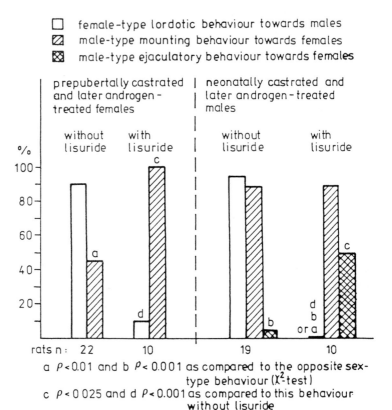

Figure 8 : Conversion of female-type and male-type sexual orientation in prepubertally castrated and later androgen-treated (0.2
mg testosterone propionate daily) female or neonatally castrated
and later androgen treated male rats induced within 1h by s.c.
injection of 0.25 mg/kg lisuride.

2. Homotypical sex hormone and/or neurotransmitter levels
   (which are characteristic for one's own genetic, gonadal and
   genital sex) organize predominantly homotypical gonado-
   trophin-secretion patterns (i.e. tonic or cyclic gonadotrophin
   secretion, non-evocability or evocability of a positive oestrogen feedback on LH secretion), predominantly homotypical

sexual orientation and/or gender role behaviour, whereas heterotypical sex hormone and/or neurotransmitter levels during brain organization, which are characteristic for the opposite sex, give rise to a predominantly heterotypical differentiation of gonadotrophin secretion patterns, sexual orientation and/or gender role behaviour (e.g. homosexuality or trans-sexuality).

3. Prenatal stress appears to be able to dedifferentiate sex-specific brain organization in both sexes.

4. 21-Hydroxylase deficiency in women represents a predisposition for female homosexuality and/or trans-sexuality.

REFERENCES

Barberino, A. and de Marinis, L. (1980). Estrogen induction of luteinizing hormone release in castrated adult human males. Journal of Clinical Endocrinology and Metabolism, 51, 280-286.

Barraclough, C.A. and Gorski, R.A. (1961). Evidence that the hypothalamus is responsible for androgen-induced sterility in the female rat. Endocrinology, 68, 68-70.

Beach, F.A. and Holz, A.M. (1946). Mating behavior in male rats castrated at various ages and injected with androgen. Journal of Experimental Zoology, 101, 91-142.

Blanchard, R. (1985). Typology of male-to-female trans-sexualism. Archives of Sexual Behavior, 14, 247-261.

Blanchard, R., Clemmensen, L.H. and Steiner, B.W. (1987). Heterosexual and homosexual gender dysphoria. Archives of Sexual Behavior, 16, 139-152.

Buhl, A.E., Norman, R.L. and Resko, J.A. (1978). Sex differences in estrogen-induced gonadotrophin release in hamsters. Biology of Reproduction, 18, 592-597.

Calne, D.B., Horowski, R., McDonald, R.J. and Wuttke, W. (1983). Lisuride and other dopamine agonists : basic mechanisms and endocrine and neurological effects, New York : Raven Press.

Dantchakoff, V. (1938). Rôles des hormones dans les manifestations des instincts sexuels. Comptes Rendus de l'Académie des Sciences, 206, 945-947.

Diamond, M. (1965). A critical evaluation of the ontogeny of human sexual behavior. Quarterly Review of Biology, 40, 147-175.

Dörner, G. (1967). Tierexperimentelle untersuchungen zur frage einer hormonellen pathogenese der homosexualität. Acta Biologica et Medica Germanica, 19, 569-584.

Dörner, G. (1969). Zur frage einer neuroendokrinen pathogenese, prophylaxe und therapie angeborener sexual deviationen. Deutsche Medizinische Wochenschrift, 94, 390-396.

Dörner, G. (1972). Sexualhormonabhängige Hehirndifferenzierung und Sexualität. Jena : VEB Fischer.

Dörner, G. (1976). Hormones and Brain Differentiation. Amsterdam : Elsevier.

Dörner, G. (1980). Sexual differentiation of the brain. Vitamins and Hormones, 38, 325-381.

Dörner, G. (1983). Prevention of impaired brain development by hormones and drugs. In : Drugs and Hormones in Brain Development, Schlumpf M. and Lichtensteiger W. (Eds.), Basel : Karger, pp. 234-240.

Dörner, G. (1988a). Sexual endocrinology and terminology in sexology. Experimental and Clinical Endocrinology, 38, 129-134.

Dörner, G. (1988b). Neuroendocrine response to estrogen in hetero-, homo- and trans-sexuals and sexual brain differentiation, Archives of Sexual Behavior, in press.

Dörner, G. and Döcke, F. (1964). Sex-specific reaction of the hypothalamo-hypophysial system of rats. Journal of Endocrinology, 30, 265-266.

Dörner, G. and Fatschel, J. (1970). Wirkungen neonatal verabreichter androgene und antiandrogene auf sexualverhalten und fertilität von rattenweibchen. Endokrinologie, 56, 29-48.

Dörner, G., Döcke, F. and Hinz, G. (1971). Paradoxical effects of estrogen on brain differentiation. Neuroendocrinology, 7, 146-155.

Dörner, G., Götz, F. and Rohde, W. (1975a). On the evocability of a positive oestrogen feedback action on LH secretion in male and female rats. Endokrinologie, 66369-372.

Dörner, G., Rohde, W. andSchnorr, D. (1975b). Evocability of a slight positive oestrogen feedback action on LH secretion in castrated and oestrogen-primed men. Endokrinologie, 66, 373-376.

Dörner, G., Rohde, W., Stahl, F., Krell, L. and Masius, W.G. (1975c). A neuroendocrine predisposition for homosexuality in men. Archives of Sexual Behavior, 4, 1-8.

Dörner, G., Rohde, W., Seidel, K., Haas, W. and Schott, G. (1976). On the evocability of a positive oestrogen feedback action on LH secretion in trans-sexual men and women. Endokrinologie, 67, 20-25.

Dörner, G., Geier, Th., Ahrens, L., Krell, L., Münx, G., Sieler, M., Kittner, E. and Müller, H. (1980). Prenatal stress as possible aetiogenetic factor of homosexuality in human males. Endokrinologie, 75, 365-368.

Dörner, G., Rohde, W., Seidel, K., Haas, W. and Schott, G. (1983a). On the LH response to oestrogen and LH-RH in trans-sexual men. Experimental and Clinical Endocrinology, 82, 257-267.

Dörner, G., Schenk, B., Schmiedel, B. and Ahrens, L. (1983b). Stressful events in prenatal life of bi- and homosexual men. Experimental and Clinical Endocrinology, 81, 88-90.

Dörner, G., Döcke, F., Götz, F., Rohde, W., Stahl, F. and Tönjes, R. (1987). Sexual differentiation of gonadotrophin secretion, sexual orientation and gender role behavior. Journal of Steroid Biochemistry, 27, 1081-1087.

Ehrhardt, A.A., Meyer-Bahlburg, H.F.L., Rosen, L.R., Feldman, J.F., Veridiano, N.P., Zimmerman, I. and McEwen, B.S. (1985). Sexual orientation after prenatal exposure to exogenous estrogen. Archives of Sexual Behavior, 14, 57-78.

Elsaesser, F. and Parvizi, N. (1979). Estrogen feedback in the pig : sexual differentiation and the effect of prenatal testosterone treatment. Biology of Reproduction, 20, 1187-1193.

Feder, H. and Whalen, R.E. (1965). Feminine behavior in neo-natally castrated and estrogen-treated male rats. Science, 147, 306-307.

Futterweit, W., Weiss, R.A. and Fagerstrom, M. (1986). Endocrine evaluation of forty female-to-male transsexuals. Archives of Sexual Behavior, 15, 69-78.

Gerall, A.A. (1967). Effects of early postnatal androgen and estrogen injections on the estrous activity cycles and mating behavior of rats. Anatomical Record, 157, 97-104.

Gerall, A.A. and Ward, F.L. (1966). Effects of prenatal exogenous androgen on the sexual behavior of the female albino rat. Journal of Comparative and Physiological Psychology, 62, 370-375.

Gladue, B.A., Green, R. and Hellman, R.E. (1984). Neuroendocrine response to estrogen and sexual orientation. Science, 225, 1496-1499.

Goh, H.H., Wong, P.C. and Ratnam, S.S. (1985). Effects of sex steroids on the positive estrogen feedback mechanism in intact women and castrated men. Journal of Clinical Endocrinology and Metabolism, 61, 1158-1164.

Gooren, L. (1986a). The neuroendocrine response of luteinizing hormone to estrogen administration in heterosexual, homosexual, and transsexual subjects. Journal of Clinical Endocrinology and Metabolism, 63, 583-588.

Gooren, L. (1986b). The neuroendocrine response of luteinizing hormone to estrogen administration in the human is not sex specific but dependent on the hormonal environment. Journal of Clinical Endocrinology and Metabolism, 63, 589-593.

Gotz, F. and Dörner, G. (1980). Homosexual behaviour in prenatally stressed male rats after castration and oestrogen treatment in adulthood. Endokrinologie, 76, 115-117.

Götz, F., Tönjes, R., Maywald, J. and Dörner, G. (1988). Influence of lisuride on sex-specific sexual behaviour in rats. Acta Endocrinologica, in press.

Goy, R.W., Bridson, W.E. and Young, W.C. (1964). The period of maximal susceptibility of the prenatal female guinea pig to masculinizing actions of testosterone propionate. Journal of Comparative and Physiological Psychology, 57, 166-174.

Goy, R.W., Phoenix, Ch.H. and Meidinger, R. (1967). Postnatal development of sensitivity to estrogen and androgen in male, female and pseudohermaphroditic guinea pigs. Anatomical Record, 157, 87-96.

Goy, R.W., Wolf, J.E. and Eisele, G. (1972). Experimental female hermaphroditism in rhesus monkeys : anatomical and psychological characteristics. In : Handbook of Sexology, Money J. and Musaph H. (Eds.), Amsterdam : Excerpta Medica, pp. 139-176.

Grady, K.L. and Phoenix, Ch.H. (1963). Hormonal determinants of mating behavior ; the display of feminine behavior by adult male rats castrated neonatally. American Zoologist, 3, 482-483.

Harris, G.W. (1964). Sex hormones, brain development and brain function. Endocrinology, 75, 627-648.

Harris, G.W. and Levine, S. (1962). Sexual differentiation of the brain and its experimental control. Journal of Physiology, 163, 42-43.

Harris, G.W. and Levine, S. (1965). Sexual differentiation of the brain and its experimental control. Journal of Physiology, 181, 379-400.

Hodges, J.K. (1980). Regulation of estrogen induced LH release in male and female marmoset monkeys (Callithrix jacchus). Journal of Reproduction and Fertility, 60, 389-398.

Hohlweg, W. (1934). Veränderungen des hypophysenvorderlappens und des ovariums nach behandlung mit grossen dosen von follikelhormon. Klinische Wochenschrift, 13, 92-95.

Hohlweg, W. and Chamorro, A. (1937). Über die luteinisierende wirkung des follikelhormons durch beeinflussung der luteogenen hypophysenvorderlappensekretion. Klinische Wochenschrift, 16, 196-197.

Imperato-McGinley, J., Peterson, R.E., Gautier, T. and Sturla, E. (1979). Male pseudohermaphroditism secondary to 5α-reductase deficiency : a model for the role of androgens in both the development of the male phenotype and the evolution of a male gender identity. Journal of Steroid Biochemistry, 11, 637-645.

Karsch, F.J., Dierschke, D.J. and Knobil, E. (1973). Sexual differentiation of pituitary function : apparent difference between primates and rodents. Science, 173, 484-486.

Kulin, H.E. and Reiter, E.O. (1976). Gonadotropin and testosterone measurement after estrogen administration to adult men, prepubertal and pubertal boys, and men with hypogonadotropism : evidence for maturation of positive feedback in the male. Pediatric Research, 10, 46-51.

McEwen, B.S., Lieberburg, I., Chaptal, C. and Krey, L.C. (1977). Aromatization. Important for sexual differentiation of the neonatal brain. Hormones and Behavior, 9, 249-263.

Müller, D., Orthner, H., Roeder, F., König, A. and Bosse, K.

(1974). Einfluss von hypothalamusläsionen auf sexualverhalten und gonadotrope funktion beim menschen : bericht über 23 fälle. In : Endocrinology of Sex, Dörner G. (Ed.), Leipzig : Johann Ambrosius Barth, pp. 80-105.

Neumann, F., von Berswordt-Wallrabe, R., Elger, W., Steinbeck, M., Hahn, J.D. and Kramer, M. (1970). Aspects of androgen-dependent events studied by antiandrogens. Recent Progress in Hormone Research, 26, 337-405.

Ohkawa, T., Arai, K., Okinaga, S., Götz, F., Stahl, F. and Dörner, G. (1986). Tyrosine administration combined with stress exposure in prenatal life prevents the diminished male copulatory behaviour in adult rats. In : Systemic Hormones, Neurotransmitters and Brain Development, Dörner G., McCann S.M. and Martini L. (Eds.), Basel : Karger, pp. 167-171.

Ohno, S. (1979). Major Sex-Determining Genes. Heidelberg : Springer.

Pfeiffer, C.A. (1936). Sexual differences of the hypophyses and their determination by the gonads. American Journal of Anatomy, 58, 195-225.

Phoenix, Ch.H., Goy, R.W., Gerall, A.A. and Young, W.C. (1959). Organizing action of prenatally administered testosterone propionate on the tissues mediating mating behavior in the female guinea pig. Endocrinology, 65, 369-382.

Pollard. I. and Dyer, S.L. (1985). Effect of stress administered during pregnancy on the development of fetal testes and their subsequentfunction in the adult rat. Journal of Endocrinology, 107, 241-245.

Roeder, F. and Müller, D. (1969). Zur stereotaktischen heilung der pädophilen homosexualität. Deutsche Medizinische Wochenschrift, 94, 408-415.

Rohde, W. (1981). Untersuchungen zur geschlechtsspezifischen ontogenese des menschen. Thesis B, Humboldt-University, Berlin.

Rohde, W., Uebelhack, R. and Dörner, G. (1986). Neuroendocrine response to oestrogen in transsexual men. In : Systemic Hormones, Neurotransmitters and Brain Development, Dörner G., McCann S.M. and Martini L. (Eds.), Basel : Karger, pp. 75-78.

Schnabl, Ch. (1983). Untersuchungen zur möglichen neuroendokrinen prädisposition der transsexualität. Thesis A, Humboldt-University, Berlin.

Stahl, F., Götz, F., Poppe, I., Amendt, P. and Dörner, G. (1978). Pre- and early postnatal testosterone levels in rat and human. In : Hormones and Brain Development, Dörner G. and Kawakami M. (Eds.), Amsterdam : Elsevier, pp. 35-41.

Steiner, A.R., Clifton, D.K., Spies, H.G. and Resko, J.A. (1976). Sexual differentiation and feedback control of luteinizing hormone secretion in the rhesus monkey. Biology of Reproduction, 15, 206-212.

Swanson, H.H., Brayshaw, J.S. and Payne, A.P. (1971). Effects of neonatal androgenization on sexual and aggressive behavior in the golden hamster. In : Endocrinology of Sex, Dörner G. (Ed.), Leipzig : Johann Ambrosius Barth, pp. 61-74.

Van de Wiele, R.L., Bogumil, F., Dyrenfurth, I., Ferin, M., Jewelewicz, R., Warren, M., Rizkallah, J. and Mikhail, G. (1970). Mechanisms regulating the menstrual cycle in women. Recent Progress in Hormone Research, 26, 63-95.

Wachtel, S.S., Ohno, S., Koo, G.C. andBoyse, E.A. (1975). Possible role for H-Y antigen in the primary determination of sex. Nature, 257, 235-236.

Wachtel, S.S., Koo, G.C. and Ohno, S. (1977). H-Y antigen and male development. In : The Testis in Normal and Infertile Man, Troen P. and Nankin H.R. (Eds.), New York : Raven Press, pp. 35-43.

Ward, I.L. (1977). Exogenous androgen activates female behavior in non-copulating, prenatally stressed male rats. Journal of Comparative and Physiological Psychology, 91, 465-471.

Westfahl, P.K., Stadelman, H.L., Horton, L.E. and Resko, J.A. (1984). Experimental induction of estradiol positive feedback in intact male monkeys : absence of inhibition by physiologic concentrations of testosterone. Biology of Reproduction, 31, 856-862.

Whalen, R.E. and Edwards, D.A. (1967). Hormonal determinants of the development of masculine and feminine behavior in male and female rats. Anatomical Record, 157, 173-180.

Whalen, R.E. and Nadler, R.D. (1963). Suppression of the development of female mating behavior by estrogen administered in infancy. Science, 141, 273-274.

Yamaji, T., Dierschke, D.J., Hotchkiss, J., Bhattacharya, A.H., Surve, A.H. and Knobil, E. (1971). Estrogen induction of LH release in the rhesus monkey. Endocrinology, 89, 1034-1041.

Young, W.C., Goy, R.W. and Phoenix, Ch.H. (1964). Hormones and sexual behavior. Science, 143, 212-218.

8 Biological and Psychological Factors in Human Aggression

Ronald Langevin *

Aggression is a complex phenomenon complicated by a host of definitional problems as well as by the wide range of biological, psychological and experimental factors that play a role in its genesis and interact with each other in subtle ways. The term "aggression" may be used in a positive sense to mean ambitious and assertive, as in "aggressive salesperson", through minor negative meanings, including irritability, argumentativeness and intrusiveness, to major negative definitions involving force, as seen in physical assaults and homicides. In order to clarify research on the topic, this chapter will examine only the extreme forms of human aggression, namely, violence such as homicide, physical assault resulting in criminal charges, and sexual assaults on strangers. Hopefully, a careful exploration of extreme cases will lead to a fuller understanding of aggression in general.

There is some consensus in modern history, where reliable statistics are available, that men are far more likely to be violent than women. Typically, 8 or 9 of 10 homicides are carried out by men, and only 1 or 2 of 10 by women (Wolfgang, 1958 ; Langevin & al., 1982a,b). Both sexes use similar means for killing and assaulting, with the use of guns being most common. Women are most inclined to kill someone they know. Whereas this is true for men as well, stranger homicides are usually male-perpetrated crimes, carried out as part of a robbery or sexual assault. Men and women differ in a great many ways, both psychologically and biologically, and some features are believed to predispose the male to greater violence.

PSYCHOLOGICAL FACTORS

History of Aggression

* Clarke Institute of Psychiatry, University of Toronto, Toronto, Canada.

Many writers have maintained that the best simple predictor of future aggression is past aggression. Megargee (1976) has noted that, in fact, past aggression may be the only satisfactory predictor of future aggression. However, when one examines what is involved in "history of aggression", a complex array of factors must be considered, such as familial violence, childhood violence, the use of physical force on others as an adult, social class, etc... Moreover, the relative weight to assign these factors in the prediction of aggression has yet to be determined. In some instances, one factor may be the significant predictor of future aggression, whereas in others a multifaceted interaction of factors results in the expression of aggressive behaviour. For example, a 15- year-old male, seen in our clinic, found, upon his sexual awakening, an overwhelming desire to kill a woman. A pampered youth from a "good family", his first sexual act was to fatally stab a stranger on the street. In such a case, aberrant libido was the single obvious factor leading to murder. However, most assaultive individuals have a history of violence that shows up in a variety of contexts. A 38-year-old heavy drinker who was receiving government relief payments was living common-law with a 35-year-old divorcee. He had been beaten as a child by an alcoholic father and he was a nervous individual as a result. He was not easily provoked to anger but he had been in bar room fights on occasion. He had a criminal record for assault and impaired driving. He frequently beat his girlfriend and when he found her in bed with his best friend, he beat her to death.

The aggressiveness of violent individuals is often difficult to detect because, as Lang & al. (1987) have shown, murderers tend to lie about their past and, in particular, to minimize the aggressiveness of their previous behaviour. Langevin & al. (1982 a, b) found that, if one used a variety of indicators, a full 70 % of men and women who kill have a previous history of aggressive behaviour which may or may not have resulted in a criminal record. Only 30 % actually had a previous criminal charge for assault.

The availability of weapons has been debated as a single most important factor in the commission of violent acts. Langevin & al. (1982a,b) showed that three times as many men and women who were convicted of homicide had weapons in their possession than did the population at large (27 % vs 9 %). More recently, Langevin & al. (1987) found that the incidence of weapon ownership was even higher, since 40 % of killers had weapons available which facilitated the commission of their violent acts. In his book Guns in America, Jervis Anderson (1984) noted that 84,644 people were murdered in the U.S.A. between 1963 and 1973, more than the 46,121 who died in the Vietnam war during the same period. The desire to possess handguns and their use are expressions of a cultural milieu so that when one examines seemingly simple factors such as "gun ownership" one is really dealing with a complex array of variables that influence the frequency with which violent acts occur.

Women often are considered to be better socialized, less aggressive and more passive than men (Cole & al., 1968). Boys are more physical in their play and are more likely than girls to respond physically to aggression. Nevertheless, girls may ostracize their antagonist or punish them in the contex of a social relationship. Thus, a boy may hit whereas a girl may tell everyone not to play with her enemy. On the other hand, females who assault and kill show a more masculine pattern of behaviour and aggressiveness. For example, McClain (1982a,b) examined 119 black female killers from 6 American cities and found 78 % were in fights as teenagers or adults, 63 % had previous legal difficulties, and 45 % grew up in homes characterized by physical fighting.

Biggers (1979) reported that 135 convicted female killers tended to see violence as a normal fact of survival. They carried out their crimes with guns or knives and felt justified in their behaviour. Interestingly, many women were determined to carry weapons "for their own protection" after their release from prison.

Family and Personal History

Part of "violence breeding violence" has its source in the family. Many offenders who are processed by the criminal justice system have a difficult and violent family background (Glueck & Glueck, 1950). In some cases the use of oppressive violence against a child results in a neurotic personality and an individual who is fearful and non-violent throughout the course of his/her life. However, substantial numbers react to violence with violence. The parent's behaviour serves as a role model so that the child learns to respond to aggression by retaliation. Contrast Mr. A., a 44-year-old obese man, facing his first criminal charge for theft, and Mr. B., a 29-year-old irritable and sarcastic male, facing his second assault charge. Both men were beaten with greater than average frequency for Canadian families. Mr. A. described his parents being "at constant war". "I tried to stay out of their way and my older sister really brought me up and kept me away from them as much as possible. But the old man would take a swing at me, usually when he was mad at mother. I still shake today when I think about it". Mr. A. stole goods from a store to supplement his income which his wife considered "inadequate". Mr. B. described his father as "a lush". "He may not come home for days on end - out on a binge. Then he'd show up when he ran out of money. He'd beat up mom, take her money and hit me when I tried to protect her. When I was fourteen, he took a swing and missed. I knocked him out. He left the next day and mom and I never saw him again". Mr. B., in a drunken state, punched a fellow drinker who insulted his mother. The police charged him with common assault.

Langevin & al. (1983) found that the family background of assaultive and homicidal men and women generally was disturbed.

They examined 109 killers, 38 non-violent offenders and 54 normal community controls using a standardized Parent Child Relations Questionnaire. They found that, although the killer's families were disturbed and aggression-filled, they did not differ from those of non-violent offenders. Antisocial behaviour results from poor and aggressive parenting in general but it does not always lead to violence on the offspring's part.

The same authors also examined a number of childhood factors presumed to predict adult violence, in particular, the triad of enuresis, firesetting and cruelty to animals. They found that the triad had little predictive value in identifying the violent criminal. Results showed that killers of both sexes were less often adopted or fostered, less often ran away from home, but that they more often stole as children. Thus aggressive familial background is an important but non-specific factor in adult violence and it is not a sufficient explanation of such behaviour.

It is interesting that the female criminal is atypical and more masculine in her pattern of aggressiveness, and possibly her aggressiveness may be a better predictor of future crime than is similar behaviour in the male.

Mental Illness

Murder is such a heinous crime that the non-professional often believes it must be carried out by mentally ill individuals. However, the incidence of mental illness among male and female killers is surprisingly low and in fact only 6 % of Canadians accused of murder and seen for psychiatric assessment, are acquitted by "reason of insanity" (Schloss & Giesbracht, 1972). Similar results have been reported elsewhere (Gillies, 1976). Case studies and uncontrolled group studies have suggested that schizophrenics, the affective disordered, and epileptics are overrepresented in the violent criminal population. This assumption has not been supported by systematic research. Langevin & al. (1982a) found that personality disorders were most common in killers and that the distribution of diagnoses was similar in killers and in non-violent psychiatric patients. Familial background and history of violence may be better predictors of violence than the presence of a mental illness. A schizophrenic individual who has violence as part of his behavioural repertoire may more readily engage in such acts when psychotic.

The antisocial personality disorder diagnosis has been linked to violent behaviour. This label is somewhat nebulous as used in North American psychiatric practice. The term may imply "general criminal tendencies" through to individuals who lack any conscience development. The reliability of this diagnostic category is poor at present and it may say no more than "killers have criminal histories".

Epilepsy has been misidentified as a factor in aggressive

behaviour. Some researchers suggest that the incidence of epilepsy in criminal populations is greater than one would expect by chance. However, this may reflect more the disposition of such individuals by a criminal justice system than a causal link between epilepsy and violence (cf. Gunn, 1969). It can be argued that epilepsy precludes violence because the behaviour, resulting in seizures, is incompatible with the coordinated and motivated behaviour required for violent acts. However, some cases have been associated with violence. In particular, temporal lobe epilepsy and its automatisms, where motor behaviour is a possibility, can lead to aggression. A 32-year-old married man, seen in our clinic, had a temporal lobe focus for his seizures. On one occasion, he took a kitchen knife and chased his wife who managed to lock him out of their apartment. A hapless neighbour was placing the key in his apartment door when the patient attacked him with the knife. Fortunately the neighbour escaped and the police arrived in time to disarm the patient. Later when the patient recovered from the seizure, he was greatly distressed but he had no memory of his aggressive behaviour. Such cases suggest that there may be some link of violent behaviour to epilepsy but they are rare and the great majority of individuals who are violent do not have epileptic seizures.

Women who are subject to postpartum psychosis may become homicidal and commit infanticide. Both infanticide and this form of psychosis are relatively rare, but present an unusual interface of social, psychological and biological factors in the crime. Often, the mother is free of symptoms for 48 hours postpartum, but then has problems with memory, is confused and may show any symptoms from the psychotic spectrum. She may kill her baby out of altruism (i.e., to save her child from the terrible world). Most cases develop within two months after delivery, but the law in a number of countries gives special status to "lactating" women who commit crimes up to two years postpartum. Little proof is required that the new mother's mind has been affected, and a plea of insanity may receive ready acceptance in court. Lukianowicz (1972), Chapman (1959) and Button & al. (1972) studied women who carried out infanticide or had obsessions of killing their children. The women were described as frustrated, unable to express anger, unusually "easy going", and with unhappy or traumatic childhoods. As Button & al. (1972) noted, the women's premorbid adjustment was borderline or severely compulsive, and biological stress, perhaps hormonally related, was an important determinant in their crime.

Personality

Research in the 1960s through the 1980s identified the "aggressive personality" as a key component in the execution of aggressive behaviour. However, the instruments used, based mainly on scales derived from the Minnesota Multiphasic Personality Inventory (MMPI), have failed to provide clear identifying personality dimensions related to violence. For example, Megargee & al. (1967) suggested that people who engage in extreme and homicidal violence overcontrol their hostility. In fact it is life's circumstances and the supppression, or repression, of aggressive feelings that results in the final explosion of extreme violence. They developed the Overcontrolled Hostility (OH) Scale and found that killers scored higher. The OH Scale has been widely used in the last 20 years. However, Langevin & al. (1982a) found that the OH Scale lacks internal consistency and, when male and female killers were examined, there was no difference in overcontrolled hostility, either on the Megargee OH Scale or in terms of reliable clinical ratings of OH. Most often, violent individuals have trouble-filled histories of personal violence and of familial violence that culminate in their criminal charges.

If aggressive personality tendencies do play a role in the commission of violent acts they are infrequent causal factors. For example, the MMPI is among the most widely used clinical and research measures in the history of psychology. The inventory has ten clinical scales sampling neuroses, psychoses and personality disorders. Four MMPI validity scales offer a check on reliability of the self-reported instrument. From the inventory profile, types are developed based on the two highest clinically significant scale elevations. These profile types have been examined for killers and assaulters. Less than 10 % of the MMPI profiles have been associated with violent behaviour, and results of these scales tend to be inconsistent in the research that has been reported (Langevin & al., 1982a).

A recent promising development for the study of personality disorders and aggression is the Milton Clinical Multiaxial Personality Inventory (Milton, 1982). The inventory samples both enduring personality traits and current clinical symptoms. Among its personality scales is narcissism, a feature reflecting inordinate self appraisal and disregard for others. Langevin & al. (1985), and Bain & al. (1988) have noted this feature in sexually aggressive men. Unpublished preliminary results showed 4 of 7 sex killers had significant elevations on narcissism. Further systematic work on the MCMI may be useful in exploring personality dimensions in violence.

Alcohol and Drug Abuse

A very important factor in the commission of violent acts is substance abuse, particularly abuse of alcohol. On the average, almost half of violent crimes are associated with drinking. The actual incidence of alcoholism varies from a low of 5 % to almost 100 %, depending on the sample and the culture (contrast McKnight & al., 1966 & Gillies, 1965). Psychiatric samples tend less often to show alcoholism, perhaps because their medication precludes its use. However, significant numbers of psychiatric patients do use alcohol.

The abuse of alcohol ranges from heavy drinking to tolerance. Heavy drinkers may more readily respond to provocation and be less inhibited in acting out their anger towards others. In fact, Wolfgang (1958) noted that homicide is often carried out because of some minor, almost trivial, altercation. For example, a 52-year-old vagrant and alcoholic brought to our clinic by the police, had killed a fellow card player because he thought he was being cheated. The killer was tolerant to alcohol and both men had been drinking steadily for days when the argument occurred. An alcoholic woman in her mid-thirties dominated her alcoholic common-law partner. They constantly abused and derided each other, and frequently threatened separation. After continuous drinking over the weekend, she bludgeoned him to death when he criticized her cooking.

Tolerance to alcohol presents a paradoxical problem in that many motor behaviours may be coordinated whereas the higher cognitive functions such as judgement and reasoning, may be faulty. Tolerant alcoholics may be able to walk a straight line for the police as they are arrested for drunken driving, even though they may show breathalizer readings of over 200 milligrams percent. Tolerance to alcohol is not well understood and it is difficult to evaluate. It may be lost when the examining professional sees the patient, three weeks to three months after incarceration, although tolerance may be evident at the time of the offence. For example, a small middle-aged woman said she had been drunk every day for at least six months prior to killing her lover. One day melded into the next, and she was not sure why she had killed her man. She was examined some time after she was sober. The time during her drinking bout was confused and her memory for it incomplete. A male in his thirties murdered a stranger after consuming at least 40 ounces of liquor and possibly more. He recalled enacting torture scenes he had seen in a movie, but was unsure of details or of his motivation for the crime. In the laboratory, he was able to walk a straight line, although blood alcohol level was over 200 mg percent (legal limit 80 mg percent).

Consumption of alcohol by the non-drinker or the social drinker, on rare occasion, produces pathological intoxication. Some individuals, after drinking only a small amount of alcohol, may be triggered into psychosis or into violent behaviour for which they have no memory (Marinacci, 1963). A 33-year-old immature and single man, who has no previous criminal record, admitted to feeling aroused by sexually sadistic acts which he never had carried out. He murdered a friend after he had consumed less than 12 ozs. of beer. He did not like the taste of alcohol and rarely had any. He found that after he had consumed the small amount of beer, he had a total memory blackout. It is during this time that the killed his friend.

The margin between tolerance and pathological intoxication is a fine one. The social or non-drinker may become pathologically intoxicated on a small amount of alcohol, whereas the tolerant drinker requires prolonged exposure to excessive amounts of alcohol to achieve similar results.

One also must consider the long-term effects of alcohol use in heavy drinkers and alcoholics, namely, mood dysphoria and sexual dysfunction (Mello & Mendelson, 1978). Killers tend to be older than assaultive individuals, and they may have more problems as a result of drinking. Mood dysphoria can result in dissatisfaction with life and/or irritability when one is sober, and thus lead to violent acts. Langevin & al. (1987) found that male killers and assaulters more often showed negative mood when sobering up from drinking alcohol than criminals who did not use alcohol excessively. Our clinical cases suggest similar results for female killers.

Alcohol leads to violence in a number of ways :
- disinhibition of an already angry individual or to give courage for premeditated murder ;
- loss of judgement and cognitive faculties in the individuals tolerant to alcohol so they are unable to evaluate the current social environment. A trivial argument or insult may lead to death ;
- pathological intoxication leading to psychosis and violence.

Alcohol abuse may be a symptom of a self-indulgent antisocial individuals who please themselves before others and who do not hesitate to use violence for their own ends.

A wide range of drugs are abused by violent individuals and by the criminal population at large. As with alcohol, pathological intoxication or tolerance may be features that lead to violent behaviour. It is not possible in this review to examine all drugs but some are particularly noteworthy for the emotions that they create. The amphetamines (speed) and phencyclidine (PCP), in particular, commonly have associated paranoid states and aggressive feelings with their use. In some cases a sudden increase in dosage of a drug can result in a psychotic state and/or the commission of violent behaviour. A 38-year-old male, seen in our clinic, was abusing amphetamines. He murdered a number of women for sexual gratification under the infuence of the drug. He was judged to be psychotic at the time of his offences. A 29-year-old women killed a roomer in her flat with no obvious motivation. She had been using amphetamines as well. She described an overwhelming feeling to injure the man who was asking for help because he was ill. The only suggestion that there may have been assaultive tendencies in this woman with no previous criminal record were the events of a few days earlier. She had been using speed and she took a stray cat into a gas station washroom and tortured the cat for a prolonged period of time but finally let it go.

In general, fewer women than men drink and, if they do, they consume less than men. Female killers, however, have a greater number of alcoholics and drug abusers in their ranks than appears in the population at large (Langevin & al., 1982b). In this respect, they are similar to male killers.

A great difficulty in evaluating the effects of street drugs on violent behaviour is that the offenders even may not know what they are using. They may believe that they are taking marijuana or LSD but the dealers may mix drugs and, in fact, there is much uncertainty as to the actual drugs that have been taken. Both alcohol and drug abuse may also be considered as biological factors that have their influence on the brain and can result in seemingly irrational and unmotivated crimes.

Situational Strains

It is a common belief that the killer may be an average citizen who is under exceptional and inescapable stressful life circumstances. Factors such as recent loss of employment, marital difficulties, separation, and divorce as well as other life stresses may result in the ultimate violence (Langevin & al., 1982b). Of course all of the foregoing stressors can interact with the use of alcohol, the familial background, etc... to result in violence. An example of a "average citizen" is of a 36-year-old man who came from a culture where family honour was paramount to one's dignity. This mild mannered man had no previous criminal record and was a steady provider for his family. His wife had tired of him and was having an affair with another man. He pleaded with her for reconciliation but she was uncompromising in her behaviour. She even brought her lover to a family event, dancing closely with him and kissing him for all to see, and humiliating her husband before everyone who was important to him. He purchased a gun and shot both his wife and her lover. A 38-year-old married woman had killed her husband of 20 years. He was a heavy drinker and beat her and the four children without reason. She had anticipated leaving him, but saw no options other than to stay, since she had no marketable job skills and, in fact, had never worked outside the home. She feared for her children, especially one evening when her husband arrived home drunk and started to smash dishes for no apparent reason. He told her to clean up before he came home. He phoned from a tavern soon after and told his wife he was going to kill them all for not having the house clean. She took his double-barrelled shotgun, waited in the dark, and killed him when he opened the door.

The "love triangle" is a significant contributor to marital violence and to homicide in North America. Individuals who find themselves having affairs may also find themselves assaulted by their marital partners. In Canada, approximately 2 in 5 homicides are linked to domestic violence (Statistics Canada, Homicide Statistics, 1979). Many of these involve the "love triangle".

Rejected spouses or lovers may retaliate with aggression equal in intensity to their infatuation with their lovers. A 39-year-old emotionally unstable woman had an affair with a married man for 10 years. He kept her involved with him by means of elaborate lies. He promised her he would leave his wife once the children were old enough and once his business was secure. When he tired of her, he purchased a gun for her because she complained of being frightened when alone. One suspects he hoped she would kill herself in her more frequently occurring depressions. When she found out he had another lover, she shot him with the gun he had bought for her. She was found not guilty by reason of insanity.

Since so many individuals survive the marital separation, divorce and romantic discord without engaging in violence, it would appear that the love triangle is a precipitating factor rather than a directly causal factor in violence. The individual who is violent under these circumstances is likely to be predisposed to violence because of other factors, e.g., alcohol abuse, family background, etc... Rarely does a single situational life stressor predict violence.

When one examines stressors systematically, their role is diminished. For example, Langevin & al. (1982b) examined the frequency of unemployment among male and female killers. Although unemployment was more than 5 times the national average, it was less in killers than it was in non-violent, criminal offenders suggesting that many individuals can survive the stress of unemployment without engaging in violence. Unemployment, like other situational stressors, if they do contribute to violence, do so only indirect.

BIOLOGICAL FACTORS

A wide range of biological factors have been linked to violent acts. However, the clear identification of many of these variables in humans has not yet been accomplished. For example, excessive levels of adrenalin may or may not be associated with violence. The role of cause and effect is difficult to ascertain, i.e., do violent individuals have higher levels of adrenalin as a result of their aggressive behaviour, or are they prompted to violence because they have intrinsically higher levels of adrenalin ? These are not easy questions to answer. Moreover, the biological assay of adrenalin as well as the measurement of its levels in the bloodstream at the time of the violent behaviour are extremely difficult if not impossible to obtain. Thus, many factors remain only hypothethically related to violent acts. Some factors, however, do appear to be more prominent in the commission of violent acts, namely, brain damage and dysfunction and endocrine abnormalities.

Brain Damage and Dysfunction

Mental retardation and structural or functional damage to the brain may result in aggressive behaviour. The mentally retarded individual may not be able readily to evaluate social norms and mores so violence may result. However, there has not been a clearly established link between mental retardation per se and violence. The mentally handicapped in general are no more likely than the population at large to be violent. Mental retardation however creates complications for disposition once it is involved as a factor in criminal charges.

Case reports in the 1950s and 1960s assumed that there are "aggression centres" in the brain (Langevin & al., 1987). Brain surgery was performed to obliterate or remove these centres in disturbed patients who failed to respond to other treatments. Based on animal studies, these operations presumed that limbic and/or temporal lobe structures (such as the amygdala) are involved in human aggressive behaviour. However, this assumption has not been supported in studies of humans. Gloor (1960), for one, noted that no case of aggression in humans has been reported as a result of electrical stimulation of the amygdala. There does not appear to be any simple aggression centre in the brain. Mark and Ervin (1970) suggested that a complex cortical stimulation pattern is necessary to trigger aggression. They presented some interesting case studies wherein they stimulated a series of electrodes implanted in the cortex of a young woman, and triggered unmotivated violent behaviour. Their interesting case reports again raise the question of some factor endogenous in the brain that causes human aggression.

Langevin & al. (1985) using Computer Tomography (CT) scans of the brain and neuropsychological tests, found that structural and functional damage to the brain was present in about a quarter of homicidal and aggressive men. However, the incidence of brain damage was not different from that found in the criminal population at large, nor were there specific brain sites associated with violence, unless it was sexualized.

Sexual aggression is no less complex than aggression in general. Some rapists are antisocial but have normal sexual needs. They take by force what they want, including sex. Other rapists have anomalous sexual preferences, such as sadism. Sadists are not aroused so much by forced intercourse as they are by the control, injury fear, terror, pleading and humiliation of their victim in their captivity. Langevin & al. (1985) and Bain & al. (1988) found that sadists, but not other sexually aggressive men, showed a disturbance in gender identity with some feminine longings, a history of violence, some hormonal abnormalities, as well as other parameters common to violent individuals, such as substance abuse. Results of CT brain scans showed dilatation of the right temporal horn of the brain in 43 % of the sadists, a finding which occurred in only 11 % of non-sadistic sexual aggressives and 13 %

of criminal controls. These authors also suggested that it is possible that sadism is more common than is recognized since 45 % of their cases were sadists. Langevin and Handy (1987) also found that stranger homicides in Canada was frequently motivated by sexual needs suggesting that the fusion of sexual and aggressive urges be more closely examined in systematic studies.

The role of brain damage and dysfunction appears to be non-specific in violence generally. Individuals who have perceptual or communication difficulties as a result of brain injury may misperceive what is occurring in their environment. They may misinterpret social cues as threats or insults when they are not so intended. For example, a patient being arrested after a charge of assault was seated in a room alone, with the door ajar. Two secretaries in the corridor, but out of view of the patient, were talking about private matters. The patient burst into the hallway, tears running down her cheeks, and yelled at them to stop making fun of her. The startled secretaries left and called the psychologist.

Some homicidal individuals are desensitized to violence and perceive their situation as less violent than, in fact, it is. For example, one 26-year-old never-married killer described his father as a "good man" who was a "social drinker" and who punished him as necessary for his behaviour. However, closer examination of this "good man" showed he was an individual who consumed a dozen bottles of beer a day and who beat his son with a belt, and punched him with his fist on a frequent basis and, at times, daily. When confronted with the actual frequency of his father's aggressive behaviour, his killer son simply said "I was a bad kid, he gave me what I deserved".

Minimal brain damage or learning disabilities may operate indirectly to influence the self-concept of the individual. An inability to read or write, poor school grades and resulting derision from other children can deeply root feelings of inadequacy and inferiority. For example, a 26-year-old twice-married, drug-abusing mother of two had one relationship after another with men who beat her. She was a slow learner, failed twice in school, and never completed high school. Having never been employed outside the home, she had no confidence to acquire job skills and left herself dependent and at the mercy of abusive men, a feature not unusual in battered women (Showalter & al., 1980). A 35-year-old reformed alcoholic, reflecting on his life of crime and violence, said, "When I realized I would never make it at school because I couldn't read, I told myself, if I can't be famous, I'll be infamous, and started to be a troublemaker".

Some organic brain syndromes are associated with aggressiveness. The capacity of afflicted individuals for sustained aggressive behaviour is quite remarkable. A-29 year-old mother accused of sexually abusing her daughter amazed a social worker with the length of time she shouted at her. A second worker timed the aggressiveness of another patient. This 46-year-old male patient shouted continuously at the examiner in our clinic for a 45-minute period, an act extremely difficult for most, even the very aggressive. Both cases were diagnosed as organic personality syndrome, a subtype of organic brain syndromes.

Brain dysfunction also may result in irritability which, in a particular individual, with the predisposing background can result in violence. A 2-year-old patient reported that he became extremely irritable when he had even one bottle of beer. He said that he felt like picking a fight for no reason at all. His previous history was of an individual who, although engaging in some antisocial behaviour, was a rather passive person. He was diagnosed as suffering from hydrocephalus.

Research on violence and brain damage and dysfunction has, for the most part, been restricted to males. It is interesting, however, that males are more susceptible to brain abnormalities than are females. Flor-Henry and his colleagues (1980, 1987, 1988) have argued that brain-related problems and abnormalities are more common in males than in females, especially in the left hemisphere, because of genetic-constitutional and developmental pathology. One can hypothesize that men are more often violent and commit more murders than women because of the higher incidence of brain damage and dysfunction, which provides the biological substrate for aggressive tendencies. At the very least, the brain is more often at a lower level of functioning in violent men and women and, in combination with alcohol or drug abuse, their brain functioning is even more reduced than it would be in men and women with normal brains. The role of the brain in violence is a complex phenomenon that has not been unravelled. Further examination of neuropsychological factors may prove fruitful in future research.

Sex Hormones

Research has long suggested that androgens, testosterone in particular, are linked not only with libido but with aggressiveness, particularly in the male. Animal research has been much more consistent in suggesting this hypothesis than human research has. However, as Bain (1987) has noted, the fact that sexual awakening in the adolescent is associated with rises in testosterone level, does link it to libidinal urges and possibly to aggression (Olweus & al., 1980).

Testosterone has been examined in violent male criminals and in volunteers with high and low scores on aggression scales such as the Buss-Durkee Hostility Inventory (Doering & al., 1975 ; Ehrenkranz & al., 1974 ; Kreuz & Rose, 1972 ; Rada, 1978 ; Mendelson & al., 1982 ; Meyer-Bahlburg & al., 1974 ; Monti & al., 1977 ; Persky & al., 1971, 1977 ; Rada & al., 1976, 1983 ; Scaramella & Brown, 1978). Research results have been inconsistent, showing either that the more aggressive individuals had higher levels of testosterone in the peripheral blood, or were no different from controls. Langevin & al. (1985) and Bain & al. (1988) noted that results based on significant marker groups such as violent offenders were more consistent in suggesting a positive association of testosterone and aggression. On the other hand, questionnaire-based studies on university students offered conflicting results. Since university students have a low propensity for violence, the studies in which they took part do not offer a fair appraisal of the testosterone hypothesis.

It is appealing to consider testosterone in examining male-female differences in aggressiveness. Males are more aggressive than females and also have higher levels of circulating testosterone in their blood stream. In the criminal population, violent men's testosterone levels are even higher than non-violent men's. Moreover, Rada & al. (1976, 1983) found that the most violent group of sexual assaulters showed increased levels of testosterone over the less violent offenders. In the sexual aggressors, testosterone increases can explain both increased libido motivating a sexual crime and increased levels of violence.

Some researchers have suggested that hormones other than androgens may be more significant. Sheard & al. (1977) found, for example, that control of LH levels in males was associated with the reduction of violent behaviour in a penitentiary population to a greater degree than testosterone was.

The results of androgen research is suggestive but not conclusive. Seldom are testosterone levels in violent men beyond those in normal males. Moreover, the results are often confounded by the role of substance abuse. Several studies have found that alcohol abuse in young men is associated with increased levels of testosterone (Rada & al., 1976, 1983 ; Bain, & al., 1988) although, in older men, one may see a decrease in testosterone from prolonged alcohol abuse. Since many violent individuals tend to be young and to be heavy drinkers, some measure of substance should be examined in endocrine studies of the violent. The results of a study by Bain & al. (1988) suggest that when alcohol abuse is controlled, there is no significant relationship between testosterone or eight other hormones, including free testosterone and aggressive tendencies.

It is still possible that some dynamic tests of endocrine functioning, i.e., levels of certain sex hormones in the brain, will show complex interactions of brain structures or damage, sex hormones, and violent behaviour.

Women are believed by some researchers to be susceptible to hormonal influence, especially as regulated by the menstrual cycle. Phases of the menstrual cycle have been linked to mood swings, accident proneness, crime, suicide and violence to others (Ribero, 1962). Dalton (1977) has researched premenstrual tension extensively, and reported that measures of hormones such as progesterone or oestrogen are not useful ; rather, the periodicity and regularity of symptoms in reference to the menstrual cycle is important. She further recommended that at least two menstrual cycles should be examined to separate the many regular cycles of life from the menstrual cycle. Since many women do not provide a reliable retrospective report of mood and behaviours at menses or before, actual daily ratings of mood and symptoms over at least two cycles is desireable. Very few studies of the menstrual cycle meet these basic criteria, and when findings on violence are considered, results are inconsistent and/or studies are uncontrolled.

The belief that violence will occur more often in the paramenstruum (four days before and four after menses), is further confounded by substance abuse in violent women. Dalton (1977) noted that alcohol excesses may be related to a premenstrual lack of self-control and depression. She described a survey of alcoholic women in which 67 % related drinking bouts to menses, with most reporting increased consumption of alcohol in the premenstruum. Thus, a similar issue arises to that for testosterone : is alcohol or a hormone most related to violence. No satisfactory research is available to answer that question. Our own unpublished retrospective data on a small sample of female killers suggested that the time of the homicide occurred randomly across the menstrual cycle.

Some endocrine disorders have been associated with illogical and erratic thinking. Parathyroid tumours or removal of the thyroids may result in illogical thinking, erratic manic-like behaviour, sexual indiscretions and, in some cases, aggressive behaviour. A 49-year-old married man was diagnosed as having a thyroid tumour that was surgically removed successfully. Prior to discovery of the tumour, his behaviour became increasingly impulsive. He would jump up and touch the ceiling, dart about the house, smash furniture and frighten his wife and children. He had little insight into his behaviour until post-surgery, when he was administered medication for his defective thyroid.

One might also expect that tumours resulting in disturbance in testosterone production may also result in irritability and aggressiveness although no clear link has been established with this phenomenon and violent acts. Just as chemically induced abnormalities can affect the brain, so can natural disease states. Any factor which can disturb brain metabolism and/or functioning may result in irrational and violent acts.

Physical Illness

A variety of physical conditions have been associated with violent tendencies, but very few appear with any noteworthy frequency. Diabetes, which may be associated with states of hypoglycaemia, can result in erratic behaviour including violence (Aldersberg & Dolger, 1939). However, diabetes per se is not necessarily linked to violence nor is hypoglycaemia. It is also noteworthy that diabetes may be a complication of alcohol abuse and a problem of causality becomes more tangled when one tries to sort out such factors.

Some individuals have biological abnormalities which have been a source of psychological tension throughout their lives and have resulted in violent acts. An individual with a long history of fighting from childhood through the time of this current incarceration at 38 years of age presented with a hare lip. He was taunted and teased by others as a child. He said that he was patient but invariably he ended up fighting at least once a week, and often more, from the first grade throughout his schooling. After a while, violent behaviour became a way of life for him. His physical disability was still a source of embarrassment to him at the time of incarceration. He has many feelings of inferiority that led to violence and, indirectly, result from his physical disability.

Any number of biological factors such as beauty, disability, skin colour, etc..., can play a role in an individual's feelings about himself and may indirectly lead to violence. Reactions to diseases may also produce hostile behaviour from the afflicted individual. Some authors have suggested that allergies may be such a factor but the evidence in this respect is far from clear (Speer, 1970).

Genetic Factors

The XYY syndrome has been associated with aggressiveness and cases of homicide have been noted (Jacobs & al., 1965). The extra Y chromosome produces higher levels of testosterone, in theory, suggesting that XYY individuals should be hyper-masculine and possibly excessively aggressive compared to the average male. The syndrome is very rare but some case reports have linked it with aggression. However, a violent predisposition due to a genetic anomaly is far from clearly elucidated. The incidence of the XYY syndrome is so small that its clear detection in a criminal population is difficult. The assumption has been made that these men will be taller than average and a systematic sample bias may have resulted from this procedure (Hook & Kim, 1971 ; Owen, 1972).

Klinefelter's (XXY) Syndrome may also be associated with violence although the theoretical genetic basis for this finding is convoluted if one relates aggressiveness to excessive testosterone levels. In both XXY and XYY groups, brain structure may differ in unknown ways. The link of the syndromes to violence is far from causal and remains a mystery.

CONCLUSION

A wide range of biological and psychological factors contribute to aggressive behaviour in humans. A few of these factors have a clearer or more definitive role than others. In clinical assessment, it is necessary to keep an open mind and do a broad examination of a range of factors from mental illness through sex hormones. If one is assuming a psychological or a biological model of aggression, the resulting equation is incomplete. Clearly an interactive model of biological and psychological factors is operative in human aggression. Some factors are more important than others for a given individual. A host of factors have been suggested as contributing to aggression but not all of these are important. Many factors have not been researched at all and much more work is required, particularly on those factors related to brain functioning and family background of individuals engaged in violent acts.

Overall, factors that affect brain functioning and endocrine functioning are more likely to be related to aggressive acts. However, these are often complicated by other factors such as substance abuse and personal history which make determination of their individual roles in violence very difficult.

Women are less likely than men to be physically aggressive but there is no simple biological or psychological explanation for this difference. Certainly the male and female brains differ, with males being more susceptible to brain damage and dysfunction. However, females are also socialized differently and are less likely to abuse alcohol and drugs which are major contributors to violence. It is interesting that violent females may be more like violent males in their socialization and substance abuse patterns. Future research on the nature of possible biological similarities of violent males and females, especially brain pathology, may be informative.

REFERENCES

Aldersberg, D. and Dolger, H. (1939). Medico-legal problems of hypoglycaemia reactions to diabetes. Annals of Internal Medicine, 12, 1804-1815.

Anderson, J. (1984). Guns in American Life. New York : Random House.

Bain, J. (1987). Hormones and sexual aggression in the male. Integrated Psychiatry, 5, 82-93.

Bain, J., Langevin, R., Dickey, R., Hucker, S.J. and Wright, P. (1988). Hormones in sexually aggressive men. Annals of Sex Research, 1, p. 63-78.

Biggers, T.A. (1979). Death by murder : a study of women murderers. Death Education, 3, 1-9.

Button, J.H., Reivitch, R.S. and Kan, L. (1972). Obsessions of infanticide : a review of 42 cases. Archives of General Psychiatry, 27, 235-240.

Chapman, A.H. (1959). Obsessions of infanticide. Archives of General Psychiatry, 1, 12-16.

Cole, K.E., Fisher, G. and Cole, S.S. (1968). Women kill : a sociological study. Archives of General Psychiatry, 19, 1-8.

Dalton, K. (1977). The Premenstrual Syndrome and Progesterone Therapy, Chicago : Yearbook Medical Publishers Inc.

Doering, C.H., Brodie, H.K.H., Kraemer, H., Moos, R.H., Becker, H.B. and Hamburg, D.A. (1975). Negative affect and plasma testosterone : A longitudinal human study. Psychosomatic Medicine, 37, 484-491.

Ehrenkranz, J., Bliss, E. and Sheard, M.H. (1974). Plasma testosterone : Correlation with aggressive behaviour and social dominance in man. Psychosomatic Medicine, 36, 469-475.

Flor-Henry, P. (1980). Cerebral aspects of the orgasmic response : normal and deviational. In Forler R. and Pasini W. (Eds.). Medical Sexology. Amsterdam : Elsevier.

Flor-Henry, P. (1987). Cerebral aspects of sexual deviation. In Wilson G.D. (Ed.). Variant Sexuality. Beckenham, Kent : Croom Helm Ltd.

Flor-Henry, P. and Lang, R. (1988). Quantitative EEG investigations of genital exhibitionists. Annals of Sex Research, 1, 49-62.

Gillies, H. (1965). Murder in the West of Scotland. British Journal of Psychiatry, 111, 1087-1094.

Gillies, H. (1976). Homicide in the West of Scotland. British Journal of Psychiatry, 128, 105-127.

Gloor, P. (1960). Amygdala. In Field J. (Ed.). Handbook of Physiology, 2, 1395-1420.

Glueck, S. and Glueck, E. (1950). Unravelling Juvenile Delinquency. Cambridge, Mass. : Harvard University Press.

Gunn, J. (1969). The prevalence of epilepsy among prisoners. Proceedings of the Royal Society of Medicine, 62, 60-62.

Hook, E.B. and Kim, D.S. (1971). Height and antisocial behaviour in XY and XYY boys. Science, 172, 284-286.

Jacobs, P.A., Brunton, M., Melville, M.M., Brittain, R.P. and McClemont, W.F. (1965). Genetics : aggressive behaviour, mental subnormality and the XYY male. Nature, 208, 1351-1352.

Kreuz, L.E. and Rose, R.M. (1972). Assessment of aggressive behaviour and plasma testosterone in young criminal population. Psychosomatic Medicine, 34, 321-332.

Lang, R., Langevin, R., Holden, R., Fiquia, N.A. and Wu, R. (1987). Personality and criminality in violent offenders. Journal of Interpersonal Violence, 2, 179-195.

Langevin, R. and Handy, L. (1987). Stranger homicide in Canada. Journal of Criminal Law & Criminology, 78, 398-429.

Langevin, R., Paitich, D., Orchard, B., Handy, L. and Russon, A. (1982a). Diagnosis of killers seen for psychiatric assessment. Acta Psychiatrica Scandanavica, 66, 216-228.

Langevin, R., Paitich, D., Orchard, B., Handy, L. and Russon, A. (1982b). The role of alcohol, drugs, suicide attemps and situational strains in homicide committed by offenders seen for psychiatric assessment. Acta Psychiatrica Scandinavica,

_66_, 229-242.

Langevin, R., Paitich, D., Orchard, B., Handy, L. and Russon, A. (1983). Childhood and family background of killers seen for psychiatric assessment. Bulletin of the American Academy of Psychiatry and Law, _11_, 331-341.

Langevin, R., Ben-Aron, M.H., Coulthard, R., Heasman, G., Purins, J.E., Handy, L., Hucker, S.J. and Russon, A.E. (1985). Sexual aggression. In Langevin R. (Ed.) Erotic Preference, Gender Identity, and Aggression in Men : New Research Studies. Hillsdale, New Jersey : Lawrence Erlbaum & Assoc., 39-76.

Langevin, R., Ben-Aron, M., Wortzman, G., Dickey, R. and Handy, L. (1987). Brain damage diagnosis, and substance abuse among violent offenders. Behavioral Science and Law, _5_, 77-94.

Lukianowicz, N. (1972). Attempted infanticide. Psychiatric Clinics, _5_, 1-16.

Marinacci, A.A. (1963). A special type of temporal lobe (psychomotor) seizure following ingestion of alcohol. Bulletin of the Los Angeles Neurological Society, _28_, 241-250.

Mark, V.H. and Ervin, F.R. (1970). Violence and the Brain. New York : Harper & Row.

McClain, P.D. (1982a). Black female homicide offenders and victims : are they from they same population ? Death Education, _6_, 265-278(a).

McClain, P.D. (1982b). Black females and lethal violence : has time changed the circumstances under which they kill ? Omega : Journal of Death and Dying, _6_, 13-25(b).

McKnight, C.K., Mohr, J.W., Quinsey, R.E. and Erochko, J. (1966). Mental illness and homicide. Canadian Psychiatric Association Journal, _11_, 91-98.

Megargee, E.I. (1976). The prediction of dangerous behaviour. Criminal Justice and Behavior, _3_, 3-22.

Megargee, E.I., Cook, D. and Mendelshohn, G. (1967). Development and validation of an MMPI scale of assaultiveness in over-controlled individuals. Journal of Abnormal Psychology, _72_, 519-528.

Mello, N.K. and Mendelson, J.H. (1978). Alcohol and human behaviour. In Iversen L.L., Iversen S.D. and Snyder S.H. (Eds.). Handbook of Psychopharmacology, _12_, 235-317.

Mendelson, J.H., Dietz, P.E. and Ellingboe, J. (1982). Postmortem plasma luteinizing hormone levels and antemortem violence. Pharmacology, Biochemistry and Behavior, _17_, 171-173.

Meyer-Bahlburg, H.F., Nat, R., Boon, D.A., Sharma, M. and Edward, J.A. (1974). Aggressiveness and testosterone measures in man. Psychosomatic Medicine, _36_, 269-274.

Millon, T. (1982). Millon Clinical Multiaxial Inventory Manual, second edition. Minneapolis, Minn : Interpretive Scoring Systems.

Monti, P.M., Brown, W.A. and Corriveau, D.P. (1977). Testosterone and components of aggressive and sexual behavior in man. American Journal of Psychiatry, 134, 692-694.

Olweus, D., Mattsson, A., Schalling, D. and Low, H. (1980). Testosterone, aggression, physical and personality dimensions in normal adolescent males. Psychosomatic Medicine, 42, 253-269.

Owen, D.R. (1972). The 47,XYY male : a review. Psychological Bulletin, 78, 209-233.

Persky, H., Smith, K.D. and Basu, G.K. (1971). Relation of psychologic measures of aggression and hostility to testosterone production in man. Psychosomatic Medicine, 33, 265-277.

Persky, H., O'Brien, C.P., Fine, E., Howard, W.J., Khan, M.A. and Beck, R.W. (1977). The effect of alcohol and smoking on testosterone function and aggression in chronic alcoholics. American Journal of Psychiatry, 134, 621-625.

Rada, R.T. (1978). Clinical Aspects of the Rapists. New York : Grune & Stratton.

Rada, R.T., Laws, D.R. and Kellner, R. (1976). Plasma testosterone levels in the rapist. Psychosomatic Medicine, 38, 257-268.

Rada, R.T., Laws, D.R., Kellner, R., Stivastava, L. and Peake, G. (1983). Plasma androgens in violent and nonviolent sex offenders. Bulletin of the American Academy of Psychiatry and the Law, 11, 149-158.

Ribero, A.L. (1962). Menstruation and crime. British Medical Journal, 1, 640.

Scaramella, T.J. and Brown, W.A. (1978). Serum testosterone and aggressiveness in hockey players. Psychosomatic Medicine, 40, 262-265.

Schloss, B. and Giesbracht, N.A. (1972). Murder in Canada. Toronto : Centre of Criminology, University of Toronto.

Sheard, M.H., Marini, J.K. and Giddings, S.S. (1977). The effect of lithium on luteinizing hormone and testosterone in man. Diseases of the Nervous System, 38, 765-769.

Showalter, C.R., Bonnie, R.J. and Roody, V. (1980). The spousal-homicide syndrome. International Journal of Law and Psychiatry, 3, 117-141.

Speer, F. (1970). Allergies of the Nervous System. Springfield, Ill : C. Thomas Publishers.

Statistics Canada, Homicide Statistics (1979). Ottawa : Ministry of Supply & Service, Cat. 85-209.

Wolfgang, M.E. (1958). Patterns in Criminal Homicide. Philadelphia, Pa. : University of Pennsylvania Press.

9 Heterotypical Behaviour in Man and Animals : Concepts and Strategies

Richard E. Whalen *

A biologist who studies aggression in mice might ask, "What can an analysis of limerence in homosexual men tell me about why castrated, but not gonadally intact, male mice attack lactating females ?" (Haug & Brain, 1978). I realize that it is more common for the student of human behaviour to ask the obverse question ; nonetheless, the question is legitimate.

The biologist fully realizes the enormous genetic and experiential distance that exists between mice and humans. Indeed, the behavioural biologist is often frustrated in attempts to extract principles of organismal physiology and behaviour on finding that members of two related strains of the same species respond differently to some biological or experiential manipulation. For example, it became evident in the 1960s that testosterone, the primary androgenic hormone secreted by the testes, was not, as long thought, the agent that regulates the growth and function of the prostate, an organ that causes medical problems, including cancer, for ageing men. Rather, testosterone is metabolized in prostatic cells to another androgen, dihydrotestosterone (DHT), and it is DHT that activates prostatic cells. A new principle emerged - gonadal hormones are active only after metabolism.

However, McDonald & al. (1970) claimed that DHT would not stimulate copulatory behaviour in castrated male rats and suggested that another metabolite of testosterone, oestradiol, normally considered an ovarian hormone, may be the "active metabolite" that controls mating behaviour in males as it does in females. Biochemists term the conversion of testosterone to oestradiol "aromatization". The Aromatization Hypothesis emerged

* University of California Riverside, Department of Psychology, Riverside, California 92521

and behavioural biologists began to proselytize for it. All behavioural functions previously thought to be regulated by testosterone were now thought to be regulated by oestradiol. Of course, evidence soon became available that DHT, which is not metabolized to an oestrogen, is fully effective in inducing copulatory behaviour in males of a variety of species as distant genetically as guinea pigs and rhesus monkeys. We found that DHT would maintain mating in castrated male hamsters (Whalen & DeBold, 1974) and in at least one strain of rat (Olsen & Whalen, 1984).

In studies of aggressive behaviour in male mice we have found one strain that is quite sensitive to DHT, another that responds readily to oestrogen and a third which is about equally sensitive to androgen and oestrogen in the hormonal control of attack against other males (Simon & Whalen, 1986).

It is very difficult to extract principles in the presence of genetic diversity. Aromatization is not a principle of the hormonal control of male-typical behaviour. It is one of several mechanisms that have emerged during evolution to regulate important adaptive functions. In this context, the biologist's question about the relevance of homosexual limerence to lactating female attack is indeed legitimate, as is the psychologist's question about the relevance of mouse behaviour to understanding human heterotypical behaviour.

It seems to me that homological thinking is unlikely to advance our understanding of either the mouse or the human. However, I would argue that analogical thinking, based upon observations of a variety of genotypes can help us to both focus our analyses and to broaden our concepts of behaviour, its functions and regulation. I would argue, for example, that a central question which links the attack of lactating mice and homosexual limerence concerns the regulation of the saliency of stimuli. Why does the homosexual male "fall in love" with an individual who displays "male" stimuli (rather than female stimuli) ? Or, to ask the even more difficult question, "Why does the homosexual respond preferentially to a subset of male stimuli ?" Or, to ask possibly the most difficult and most general question, what factors determine individual differences in our preferences for "falling in love" stimuli ? Buss (1989) has recently suggested that there may even be human "universals" in the nature of "partner stimuli" that are differentially preferred by men and women.

The value of these questions for the mouse-orientated biologist is that they will lead him or her to put on the back burner, that is, defer, rather than reject, one of two major determinants of stimulus saliency, stimulus detectability. While it may be true in baseball that you can't hit what you can't see, it is not the case that the homosexual who finds young, thin, hairy men more erotically exciting than older, fatter and balder men cannot detect the characteristics of the latter. In fact, it is his ability to make such sensory discriminations that leads to his erotic preference.

With this understanding, the mouse biologist might defer consideration of the possibility that castration of his mouse leads to a major change in the mouse's sensory detection threshold for the critical stimuli, presumably odorants, that are carried by the lactating mouse. Although hormones do modulate sensory thresholds (Parlee, 1983), he would defer studies on the effects of castration or hormone administration on sensory function and begin to ask the enormously more difficult questions about the ways in which hormones regulate the meaning of stimuli, that is, of perception.

In this instance, analogical, but not homological thinking would guide the mouse biologist. Every so often perfume makers try to convince us that they have a product, developed as a result of animal biological research, which is guaranteed to excite erotic interests in the opposite sex. This is homological thinking based on our understanding that for some mammals odorants play important roles in regulating sex and aggression. Humans, however, are dominated more by visual than olfactory stimuli. Lingerie makers succeed in their quiet way while perfumerers continue to struggle to find that guaranteed erotic attractant. Our mouse biologist should not waste a moment considering the relative visual qualities of the lactating and non-lactating females that he presents to his hormone-stimulated and hormone-free males. Thinking homologically from human to mice seems an ill-advised strategy for the biologist ; thinking analogically, however, could lead the biologist to the interesting questions about the regulation of heterotypical aggressive behaviour.

SEX DIFFERENCES

The title of this book must cause us to focus on sex diffe-rences in behaviour. There is a rich biological literature on sex differences which long predates our contemporary scientific thinking. The ancient Greeks knew well that one approached the bull and the cow differently. And they knew the benefits of, if not the reason why, castration of the bull makes for a more placid animal. They even knew something of sex differences in human behaviour. They knew, for example, that some women would show clear changes in their behaviour at periodic intervals, changes that we now associate with the premenstrual syndrome. And they knew that some men preferred to behave as women and incorporated that concept into their culture. However, it is unlikely that our ancestors were as concerned about human sex differences as we are today.

One of the important messages brought to us by Hurtig and Pichevin (this volume) is that our current concern with human sex differences has tended to minimize and obscure intrasex variability and, equally importantly, intersex similarities in behaviour. Their notion of "typicality" emphasizes the similarities between the behaviour of men and women.

Students of animal behaviour also dichotomize as male and female when typicality might be a more accurate description. For example, as reviewed by Goy and Roy (this volume), females of several mammalian species mount other females and even males in what we consider a "male-like" fashion. Thus, mounting, often with pelvic thrusting, should not be considered a male behaviour, but merely a behaviour which occurs with a higher probability and frequency in individuals with testes rather than ovaries.

As stated by Hurtig and Pichevin, "The psychology of sex differences has often gone all too quickly from sex-related differences... from the prototypical to the typical, from a grouping variable to a causal variable". Both animal behaviourists and human psychologists might heed that warning.

Goy and Roy also cite research by Hsu and Carter (1986) which reveals an interesting parallelism with the research of Eagly and Steffen (1986) cited by Hurtig and Pichevin. A number of investigators, including this author (DeBold & Whalen, 1975) have reported that the female hamster, unlike the female rat, does not show male-like mounting even when given testosterone in adulthood. Hsu and Carter, however, have shown that group-housed female hamsters show frequent mounting responses. Thus, as with the expression of aggressive behaviour by women, the social setting has a profound effect upon the nature and intensity of heterotypical activity.

Individual Differences. As the science of behaviour analysis has progressed, statistical sophistication has progressed in concert. This has had the positive result of reducing errors of generalization based upon idiosyncratic observations, no matter how dramatic. Dramatic observations now lead to further studies and not to the promulgation of elaborate theories.

On the negative side, statistical sophistication has also led to journal reporting of means and variances, along with their t-test and F-test significances to the exclusion of considerations of the nature of the variability inherent in the data. This practice has even been carried to an absurd extreme. Weiss and Coughlin (1979) published a paper with, to me, the intriguing title "Maintained aggressive behaviour in gonadectomized male Siamese fighting fish". This paper presents no mean scores and no variances. A single table contains only F-scores, degrees and freedom and "p" values.

Hurtig and Pichevin remind us that a few individuals who exhibit extreme scores on some measured variable can inordinately influence average group differences between the sexes. One must agree that "The shape of the distribution and the number of cases in each sex category should thus be studied systematically...".

In my own field, the textbook generalization is that castrated male mice stop fighting. As we recently reported (Whalen & Johnson, 1987) this is not at all true. Following castration, male mice become more variable in their attack behaviour against other males, but they certainly do not cease to be aggressive. Moreover, as reviewed by Haug and Brain (1978), castrated male mice begin attacking lactating females. Total attack frequency against male and lactating female stimuli may change little with castration in mice, but the target of that aggression changes dramatically.

Attention to individual differences, and to, as Hurtig and Pichevin say "the shape of distributions" is indeed important in studies of behaviour.

SEXUAL DIFFERENTIATION

The term "sex differences" pertains to measured differences within the two categories, male and female, that we normally employ. The term "sexual differentiation" refers to the processes that lead to sex differences. The primary determinant of sexual differentiation studied by animal scientists is hormonal. The student of human behaviour must consider not only the hormonal determinants of differentiation (see Money and Götz & al., this volume), but also the social and cultural pressures toward "sex conformity".

The single study that stimulated a flood of research on the sexual differentiation of animal sexual behaviour was that of Phoenix & al., in 1959. Those workers showed that the female offspring of guinea pigs given testosterone during pregnancy showed enhanced levels of male-like mounting when testosterone-treated in adulthood. They proposed that the androgen, acting during foetal development, "organized" the brain in the male direction and this resulted in "masculinized" or heterotypical behaviour.

In the years that followed, several investigators, using a variety of species, demonstrated that androgen treatment of developing foetuses or neonates could make females male-like in their behavioural potential. We, for example, reported that female rats testosterone-treated both prenatally and neonatally would, when given testosterone in adulthood, show mount, intromission and multiple ejaculation-like responses which were indistinguishable in their frequency and timing from those of male rats tested to the point of sexual satiation (Whalen & Robertson, 1968).

Clemens (1974) was the first to provide evidence that the effects of experimental androgenization mimic naturally-occurring events. As mentioned, it is not uncommon for female rats to exhibit mounting responses. Clemens found that the propensity of female rats to mount depended upon the sex of their neighbours in utero. Females situated between two males in utero (2M females) showed substantial levels of mounting in adulthood. Females which developed between two females (OM females) did not.

Gandelman & al., (1977) extended these observations to aggressive behaviour in mice. Females which had developed between two males were much more sensitive to androgen treatment in adulthood than were females which had developed only in proximity to other females with respect to attack directed against intruder males. Vom Saal has extended that research substantially and reviews it in his chapter.

Goy and co-workers extended the research on experimental androgenization to rhesus monkeys and in so doing revealed at least two conceptually important phenomena. First, their prenatally androgenized rhesus females showed increased frequencies of male-like mounting (and male-like play behaviours), but, more importantly, these behaviours occurred in the absence of further gonadal hormone stimulation. In the rodent studies mentioned earlier, androgen treatment in adulthood was needed to reveal the masculinizing action of hormones given early in development. In the monkeys, prenatal androgenization induced gonadal hormone-independent male-typical behaviours.

A second important facet of this research was the finding that androgenized monkeys would show male-typical partner preferences. Partner preferences are, of course, a notable feature of human erotic behaviour and of limerence. Partner preferences are also a feature of the sexual behaviour of dogs and have been observed in cats. These studies of monkeys, however, are the first to suggest that stimulus selection may be another result of hormonal stimulation during early development. If now we could only learn the nature of the mechanisms or processes which mediate stimulus saliency, we might be able to make more rapid progress toward understanding not only heterotypical behaviour in animals, but heterotypical behaviour in humans as well.

THE BRAIN : THE ORGAN OF BEHAVIOUR

The paper by Phoenix (1959) which so excited researchers and stimulated experiments on behavioural sex differentiation in animals is surprisingly circumspect about brain differentiation. Their model was the role of gonadal hormones on the differentiation of the genital tract. They did not seem to expect to find in the brain anything of the magnitude of the effect of androgens on the development of the penis. The findings of Raisman and Field (1973) that the presence of androgens during the sensitive period of development alters the relative frequency of "spine" and "shaft" synapses in the preoptic area of rats, a relatively subtle effect, might have been in line with their expectations.

However, beginning in 1978, Gorski and colleagues began to report gross morphological differences between the brains of male and female rats in the region of the preoptic area and that these differences were the result of gonadal hormone action during development (Arnold & Gorski, 1984). A little over a decade later, Allen & al. (1989), provided evidence that the preoptic area of men and women also differ morphologically.

My own research on rats suggested that the primary difference between males and females was that males were unresponsive to the lordosis inducing properties of oestrogen (Whalen & Edwards, 1967). (As an aside, but one that reminds us that genotype must always be borne in mind, European male rats are much more likely than are American rats to show lordosis when given oestrogen ; cf. Aron & al., this volume, and Whalen & al., 1969). We used radioactive tracer techniques to detect sex differences in the cellular response to oestrogen. After several frustrating studies which showed no sex difference in gross brain response to oestradiol (e.g., Whalen & Luttge, 1970) we found that hypothalamic nuclei (Whalen & Massicci, 1975) and nuclear chromatin (Whalen & Olsen, 1978) bound oestradiol less avidly in males than females and that the magnitude of hypothalamic nuclear chromatin binding of oestrogen was strongly influenced by hormones acting during development (Olsen & Whalen, 1980). Thus, the brains of males and females differentiate under the influence of hormones to an extent undreamed of by Phoenix & al. (1959).

Correlates and Causes. The discovery of gross morphological sex differences in the brain opened the possibility that we might soon understand the anatomical basis of sex differences in behaviour. That promise has been unfulfilled. To cite one example, Gorski & al. (1978) discovered a prominent sex difference in a subregion of the preoptic area, now called the sexually dimorphic nucleus of the preoptic area (SDN-POA) which is larger in males than females. Arendash and Gorski (1983), however, found no change in sexual behaviour in male rats following SDN lesions.

The complexity of the problem of relating sex differences in brain morphology to sex differences in behaviour can be illustrated by two questions :
1. Which sex differences in morphology should be related to behaviour ? Although the SDN-POA is several-fold larger in male than female rats, the more extended region in which the SDN-POA is imbedded, termed the SDNC-MPAH (sexually dimorphic nuclear complex of the medial preoptic-anterior hypothalamic area) is larger in female than male rats, hamsters, mice and guinea pigs (Bleier & al., 1982).
2. Which sexually differentiated behaviour should be related to sex differences in morphology ? Bleier & al. (1982) found that the SDNC-MPAH is larger in female than male mice. Male and 2M female mice are more sensitive to testosterone-induced attack against male stimuli than are OM females. Indeed, Gandelman

and Graham (1986) have recently reported that female mice which are the sole uterine residents during gestation (these "singleton" females are produced surgically) do not attack intruder males even when given testosterone chronically during adulthood. Yet, as reviewed by Vom Saal (this volume) OM males are more sexually active than are 2M males. Since it is also the case that there is no simple relationship between the male genotype and aggression in mice (Haug & al., 1986), it becomes clear that we have much to learn about the relationship between brain morphology and behaviour. This is particularly true for the human species for which we know so little about the functions of subcortical structures.

One Level Lower. Another question that we can ask is how brain differences could lead to behavioural differences. Our own research led us to believe that male rats do not respond to oestrogen because of permanent structural changes in nuclear chromatin that prevent an effective interaction between the intracellular protein receptors that initially bind oestrogen and the DNA-containing chromatin. However, research in several laboratories has now shown that male rats can respond to oestrogen with the display of lordosis. For example, Hennessey & al. (1986) reported that lesions of the preoptic area of male rats allow lordotic responding and Yamanouchi and Arai (1978) found that knife cuts that separate the septum from the preoptic area ("anterior roof deafferentiation" it is called) also lead to hormone-induced lordotic responding in male rats.

Aron and colleagues (this volume) review and expand upon these findings and relate components of the amygdaloid complex of the limbic system to behaviour control in both male and female rats. One feature of their research that is particularly fascinating and, I believe, conceptually important, concerns the relationship of the olfactory system to lordotic behaviour in both sexes.

One particularly exciting experiment merits comment. Aron and colleagues castrated adult male rats and all were given oestrogen. Some males were also exposed to the odour of male urine. This had no effect upon their tendency to display female-typical lordosis behaviour. Some males were also given progesterone following oestrogen "priming". This also had little effect on lordotic responding. However, if these males were given oestrogen and progesterone and exposed to the urine of other males, they displayed lordosis. There was a clear interaction between stimulation by a particular hormone, progesterone, and the response to the sensory stimulus, the odour of male urine, in the regulation of heterotypical behaviour.

I believe that a current major challenge to both animal and human psychobiologists is the development of a conceptual framework that will guide studies on the regulation of stimulus saliency and the role of perception in the regulation of hetero- and homotypical behaviour. In this context, I will speculate that sexually differentiated brain nuclear areas are involved not in the control of motor responses such as mounting and lordosis, but in the regulation of the meaning and saliency of those stimuli that mediate behaviour.

HUMAN BEHAVIOUR

I have been studying rodent sexual and aggressive behaviours for many years. I hope that some useful concepts, if not universal principles, have emerged from that research. I have always admired my colleagues who have chosen the human and other primate species as their subjects. It has seemed to me that primates present (at least) three formidable problems for the experimentalist. The first, of course, is that you cannot either ethically or practically carry out biological or experiential manipulations that are justifiable and practical with primates. The second concerns life-span. One must be very confident that some manipulation made early in the life-span of a primate will yield a theoretically important outcome if that outcome is not revealed for five, ten, twenty or more years. The third is the problem of conformity whether it be gender conformity (Hurtig & Pichevin, this volume) or cultural conformity. You are always faced with the question of whether your treatment yielded a "pure" effect, that is, one that would have been revealed regardless of the social context in which your subject developed or whether your outcome was social context specific. Of course, one can utilize "natural" experiments to study human behaviour. Money (this volume) has shown us how powerful this approach can be. Throughout his research career, Money has studied individuals who are atypical biologically either for genetic or endocrinological reasons in order to isolate those factors which contribute to human erotosexual activity and Gender Identity/Role. His insight into the hetero-sexual/homosexual dichotomy has been particularly valuable. In his chapter he reminds us of important distinctions. For example, he notes that limerence or falling-in-love is not a sex or G-I/R differentiated process any more than are the physical sexual acts engaged in by heterosexuals or homosexuals. Money also reminds us that the erotic and non-erotic components of sex role and identity emerge in individuals as epigenetic phenomena. Homosexual behaviour is not simply the result of an atypical hormonal environment prior to birth nor the simple result of rearing practices. Heterotypical and homotypical states are the result of "multivariate complexity". His discussion of individuals with 5-alpha reductase deficiency is pertinent in this regard.

Götz and colleagues (this volume), in contrast, approach G-I/R from a quite different perspective, one that I termed "homological" earlier. This research programme, now extending consistently, used sexual differentiation in the rat as a model for the development of human gender orientation. Since this approach has been critically evaluated by others (e.g., Beach, 1979) I will not pursue that discussion. What I have found exciting, as an experimentalist, from this programme is the use of the "challenge" experiment to test neuroendocrine reactivity. Heterosexuals and homosexuals are given exogenous oestrogen and one follows changes in the secretion of gonadotrophins and gonadal hormones. There have been variants on this experiment, and while the results have not always been consistent, the approach is novel and valuable. If now we could determine where the brain, the pituitary, the gonads and exogenous hormones act to alter neuroendocrine responses, we might further our understanding of gender differences in biological reactivity.

The Götz & al. review raises one additional problem we have in trying to understand human behaviour : individual history. For some time Dörner's group has been interested in the role of prenatal stress in gender development. They have, for example, studied individuals who were born in Germany during World War II. The problem, unfortunately, is that one can never determine whether those individuals studied were unusually "stressed" by the events of the war or whether other, non-stressful, but war-related factors contributed to their later behaviour.

Langevin (this volume) is well aware of the problems of individual history in his analyses of human aggression. It has been difficult to relate the human propensity for violence to hormonal, familial or personal history.

Human violence and human erotic behaviour present a very curious contrast. Violence is often associated with the ingestion of drugs, alcohol in earlier days, today cocaine. While individuals over recorded history have sought aphrodisiacs without much success, over recorded history they have had available alcohol, a drug which clearly increases the probability (releases ?) of violent behaviour. Why sex and aggression have such different susceptibilities to pharmacological control might be worth pondering, but more importantly in the present context, it would be useful to have more information on whether the two sexes are equally responsive to the violence-inducing properties of this common drug.

Langevin provides us with little support for the "raging hormone" hypothesis of human violence. Although it is clear that men tend to be more violent than women and that this violence emerges, as does erotic desire, at the time of puberty, Langevin also makes it clear that violence is certainly not unknown amongst women. He also makes it clear that "multivariable complexity" characterizes human violence as it does human sexuality.

MODELS AND STRATEGIES

To return to our biologist who asks the relevance of human limerence to the regulation of attack behaviour in mice, I will conclude that there is indeed relevance. However the questions that one must ask must be at a higher conceptual level than that of particular mechanisms. There is no reason to think that the function of the human hypothalamus mimics the function of the rat hypothalamus. There is no reason to think that "falling-in-love" is at all relevant to the copulatory behaviour of rats or that lordosis is in any way a model for sexuality in women. However, analogical thinking about more general issues, such as the regulation of stimulus saliency, the control of interindividual communication, the processes that regulate the development and decline of behaviour over the lifespan, distinctions between hormone-regulated and hormone-independent behaviours, all issues which can exist at a conceptual level that allows us to move across species, should further our attempt to develop theories of behaviour and strategies for their refutation.

REFERENCES

Allen, L.S., Hines, M., Shryne, J.E. and Gorski, R.A. (1989). Two sexually dimorphic cell groups in the human brain. Journal of Neuroscience, 9, 497-506.

Arendash, G.W. and Gorski, R.W. (1983). Effects of discrete lesions of the sexually dimorphic nucleus of the preoptic area and other medial preoptic regions on the sexual behavior of male rats. Brain Research Bulletin, 10, 147-154.

Arnold, A.P. and Gorski, R.A. (1984). Gonadal steroid induction of structural sex differences in the central nervous system. Annual Review of Neuroscience, 7, 413-442.

Beach, F.A. (1979). Animal models of human sexuality. In Sex, Hormones and Behaviour, Ciba Foundation Symposium No. 62, Excerpta Medica, pp. 113-143.

Bleier, R., Byne, W. and Siggelkow, I. (1982). Cytoarchitectonic sexual dimorphisms of the medial preoptic and anterior hypothalamic areas in guinea pig, rat, hamster and mouse. Journal of Comparative Neurology, 212, 118-130.

Buss, D.M. (1989). Sex differences in human mate preferences : Evolutionary hypotheses tested in 37 cultures. Behavioral and Brain Sciences, 12, 1-49.

Clemens, L.G. (1974). Neurohormonal control of male sexual behavior. In Reproductive Behavior, W. Montagna and W.A. Sadler (Eds), Plenum : N.Y. , pp. 23-53.

DeBold, J.F. and Whalen, R.E. (1975). Differential sensitivity of mounting and lordosis control systems to early androgen treatment in male and female hamsters. Hormones and Behavior, 6, 197-209.

Eagly, A.H. and Steffen, V.J. (1986). Gender and aggressive behavior : a meta-analytic review of the social psychological literature. Psychological Bulletin, 100, 309-330.

Gandelman, R. and Graham, S. (1986). Singleton female mouse fetuses are subsequently unresponsive to the aggression-activating property of testosterone. Physiology and Behavior, 37, 465-467.

Gandelman, R., Vom Saal, F.S. and Reinisch, J.M. (1977). Contiguity to male foetuses affects morphology and behaviour of female mice. Nature, 266, 722-724.

Gorski, R.A., Gordon, J.H., Shryne, J.E. and Southam, A.M. (1978). Evidence for a morphological sex difference within the medial preoptic area of the brain. Brain Research, 148, 333-346.

Haug, M. and Brain, P.F. (1978). Attack directed by groups of castrated male mice towards lactating or non-lactating intruders : a urine-dependent phenomenon. Physiology and Behavior, 21, 549-552.

Haug , M., Brain, P.F. and Alias Bin Kamis (1986). A brief review comparing the effects of sex steroids on two forms of aggression in laboratory mice. Neuroscience and Biobehavioral Reviews, 10, 463-468.

Hennessey, A.C., Wallen, K. and Edwards, D.A. (1986). Preoptic lesions increase the display of lordosis by male rats. Brain Research, 370, 21-28.

Hsu, C.H. and Carter, C.S. (1986). Social isolation inhibits male-like sexual behavior in female hamsters. Behavioral and Neural Biology, 46, 242-247.

McDonald, P., Beyer, C., Newton, F., Brien, B., Baker, R., Tan, H.S., Sampson, C., Kitching, P., Greenhill, R. and Pritchard, D. (1970). Failure of 5α-dihydrotestosterone to initiate sexual behaviour in the castrated male rat. Nature, 227, 964-965.

Olsen, K.L. and Whalen, R.E. (1980). Sexual differentiation of the brain : Effects on mating behavior and (3H)-estradiol binding by hypothalamic chromatin in rats. Biology of Reproduction, 22, 1068-1072.

Olsen, K.L. and Whalen, R.E. (1984). Dihydrotestosterone activates male mating behavior in castrated King-Holtzman rats. Hormones and Behavior, 18, 380-392.

Parlee, M.B. (1983). Menstrual rhythms in sensory processes : A review of fluctuations in vision, olfaction, audition, taste and touch. Psychological Record, 93, 539-548.

Phoenix, C.H., Goy, R.W., Gerall, A.A. and Young, W.C. (1959). Organizing action of prenatally administered testosterone propionate on the tissues mediating mating behavior in the female guinea pig. Endocrinology, 65, 369-382.

Raisman, G. and Field, P.M. (1973). Sexual dimorphism in the neuropil of the preoptic area of the rat and its dependence on neonatal androgen. Brain Research, 54, 1-29.

Simon, N.G. and Whalen, R.E. (1986). Hormonal regulation of aggression : Evidence for a relationship among genotype, receptor binding and behavioral sensitivity to androgen and estrogen. Aggressive Behavior, 12, 255-266.

Weiss, C. and Coughlin, J.P. (1979). Maintained aggressive behavior in gonadectomized male Siamese fighting fish (Betta splendens). Physiology and Behavior, 23, 173-177.

Whalen, R.E. and DeBold, J.F. (1974). Comparative effectiveness of testosterone, androstenedione and dihydrotestosterone in maintaining mating behavior in the castrated male hamster. Endocrinology, 95, 1674-1679.

Whalen, R.E. and Edwards, D.A. (1967). Hormonal determinants of the development of masculine and feminine behavior in male and female rats. Anatomical Record, 157, 173-180.

Whalen, R.E. and Johnson, F. (1987). Individual differences in the attack behavior of male mice : A function of attack stimulus and hormonal state. Hormones and Behavior, 21, 223-233.

Whalen, R.E. and Luttge, W.G. (1970). Long-term retention of tritiated estradiol in brain and peripheral tissues of male and female rats. Neuroendocrinology, 6, 255-263.

Whalen, R.E. and Massicci, J. (1975). Subcellular analysis of the accumulation of estrogen by the brain of male and female rats. Brain Research, 89, 255-264.

Whalen, R.E. and Olsen, K.L. (1978). Chromatin binding of estradiol in the hypothalamus and cortex of male and female rats. Brain Research, 152, 121-131.

Whalen, R.E. and Robertson, R.T. (1968). Sexual exhaustion and recovery of masculine copulatory behavior in virilized female rats. Psychonomic Science, 11, 319-320.

Whalen, R.E., Edwards, D.A., Luttge, W.G. and Robertson, R.T. (1969). Early androgen treatment and sexual behavior in female rats. Physiology and Behavior, 4, 33-39.

Yamanouchi, K. and Arai, Y. (1978). Lordosis behaviour in male rats : Effect of deafferentation in the preoptic area and hypothalamus. Journal of Endocrinology, 76, 381-382.

SUBJECT INDEX